Peer Instruction for Astronomy

Paul J. Green
Harvard University

ei PRENTICE HALL SERIES IN EDUCATIONAL INNOVATION

Prentice
Hall

Pearson Education, Inc.
Upper Saddle River, New Jersey 07458

Senior Editor: *Erik Fahlgren*
Assistant Editor: *Christian Botting*
Production Editor: *Donna Young*
Assistant Managing Editor, Science Media: *Nicole Bush*
Manufacturing Manager: *Trudy Pisciotti*
Manufacturing Buyer: *Alan Fischer*
Copy Editor: *Luana Richards*
Art Director: *Jayne Conte*
Cover Design: *DeFranco Design, Inc.*
Cover Image: *WIYN/NOAO/NSF*

Printed in the United States of America
10 9 8 7 6 5 4 3 2 1

ISBN 0-13-026310-9

Pearson Education LTD., *London*
Pearson Education Australia PTY. Limited, *Sydney*
Pearson Education Singapore, Pte. Ltd.
Pearson Education North Asia Ltd, *Hong Kong*
Pearson Education Canada, Ltd., *Toronto*
Pearson Educacion de Mexico, S.A. de C.V.
Pearson Education—Japan, *Tokyo*
Pearson Education Malaysia, Pte. Ltd.

CONTENTS

Foreword

by Eric Mazur

Interactive teaching is nothing new. After all, the great ancient Greek philosopher Socrates taught by questioning, not by telling. Still, some 25 centuries later, most science courses are still taught in a passive, expository mode. Even though many studies have shown that the traditional approach to teaching science fails and that interactive teaching methods more than double gains in understanding, most instructors in science courses continue to use non-interactive, one-way methods of presentation in their classes. I can think of a number of reasons for this tenacity of the traditional method. One reason is that changing from a passive to an interactive teaching method requires the development of new material, specifically, appropriate questions to stimulate thinking and evaluation. Another reason is that change requires adjustment in attitude and expectation by the instructor, by the students, and by colleagues.

Fortunately, numerous articles in professional journals and the public press have begun to focus attention on the need for change. With *Peer Instruction for Astronomy,* Paul Green provides a ready-to-use set of ConcepTests that will make it easier for instructors to switch from a passive lecture format to an interactive, collaborative teaching method conducive to learning and creative thinking. A recent survey of hundreds of introductory science teachers by my group has shown that the use of ConcepTests is widespread and that over 90% of instructors who adopt the method consider their implementation to be a success and plan on continuing to teach interactively. While most of the instructors are in physics departments where Peer Instruction and ConcepTests were first introduced, some 18% of respondents to the survey are from other departments, such as chemistry, mathematics, life sciences, astronomy, and engineering. It is clear from the survey that the method is popular with both students and faculty and that the demonstrated and documented improvements in learning are helping instructors convince their colleagues of the effectiveness of the method.

Sets of ConcepTests are available in print for Physics and Chemistry, and with this book Paul Green adds Astronomy to the list. A book with ConcepTests for Calculus is in production. The questions presented here cover the wide range of topics found in a typical introductory astronomy course and has enough variety to be used at various class levels within and across institutions.

Paul has facilitated the use of ConcepTests in astronomy since 1998 by making hundreds of questions available over the Web. Over 30 collaborators contributed to the wide variety and high quality of the astronomy ConcepTests. Paul has also facilitated the implementation of Peer Instruction with easy-to-print transparencies of the ConcepTests, printable flashcards for students, and on-line modifiable forms.

With this book Paul Green has removed the largest obstacle in implementing interactive teaching in introductory astronomy classes. His questions will help you turn passive students into active learners with a renewed sense of appreciation for astronomy.

Preface

The most beautiful thing we can experience is the mysterious.
It is the source of all true art and science.
Albert Einstein (1879–1955)

A primary attraction of astronomy is that it provides a sense of the mystery and majesty of the Universe. The delights of astronomy are obvious even to the uninitiated. Every child wonders about the stars and planets, about where we came from, and where we might go. For many people, astronomy is a life-long interest, fueled by the rush of discoveries emanating from the astronomical community. Every day astronomers both professional and amateur, using telescopes in space and on the ground, are unveiling the spectacular dramas of the creation and evolution of the cosmos. Besides hosting a wild menagerie of exotic phenomena, the Universe now seems to be pervaded by dark matter, and dark energy. Theorists are probing back to the Big Bang, and far into the future toward the ultimate fate of the Universe.

Few people can escape the intrigue being communicated by astronomers and the media to the public. So while astronomy is a window to the Universe, it is also one of science's primary portals to the public imagination. Many undergraduates never enroll in any science course other than astronomy, and introductory astronomy is a common course for many science majors. For those who seek a deeper understanding, the most common college-level introduction of all is the introductory college astronomy course. This is also one of science educators' best chances to emphasize the methods and relevance of scientific inquiry. Astronomy education thus provides an important boost to scientific literacy — the public understanding and enjoyment of science. Public support and the long-term health of the discipline of astronomy also depends on good astronomy education.

Counting only degree-granting physics and astronomy institutions, the AIP reported in 1994–95 a combined introductory astronomy enrollment of about 155,000 (Fraknoi 1998). A crude projection to include all 4-year colleges and 2-year colleges boosts the number to about 230,000 astronomy students each year (U.S. Census Bureau 2000). Nevertheless, the huge range of distance, size, and mass scales often presents particularly challenging conceptual problems to introductory astronomy students, even those with substantial science and mathematics preparation. The incredible diversity of astrophysical phenomena, from shooting stars to the cosmic microwave background, in all their exotic splendor, seems to defy a unified understanding.   But it's not just the exotic that escapes people. In 1988 the Public Opinion Laboratory at Northern Illinois University surveyed 2,041 American adults to get a sense of their scientific literacy.   When asked whether the Earth goes around the Sun or the Sun around the Earth, 21% got it wrong and 7% said they did not know.   Of the 72% who answered correctly, 45% said one year, 17% said one day, 2% said one month, 8% said they did not know (Fraknoi 1998).

Educational innovations that have been shown to improve learning are crucial to boost students' scientific literacy, to enhance their understanding and enjoyment of science, and to prepare them for a rich and rewarding life in the Renaissance that is the 21st century. Peer Instruction is such an innovation. Sharing ideas and engaging in debate with their peers has proven to be an engaging and effective method for students to begin to understand and appreciate the breadth of the Universe.   Under your direction as instructor, students personally confront the conceptual puzzles of astronomy, while you get the chance to assess their comprehension in real-time.

A substantial library of conceptual multiple-choice questions, or "ConcepTests" provided in this book in Chapter Five, and on CD-ROM, constitutes a primary tool for the implementation of *Peer Instruction for Astronomy*, and forms the bulk of this book. The ConcepTest Library includes problems that address many common misconceptions. Instructors are encouraged to contribute more to this Library, to make it even more useful. You are also invited to share your thoughts and experiences, either informally or as part of more formal assessment efforts being launched to evaluate the particular challenges and benefits of applying Peer Instruction to undergraduate astronomy. By helping students replace the shaky struts and oddly tilted pillars in the foundations of their knowledge, by helping them develop and use a tool chest of physical concepts grounded in their own experience, we can help them to build a stronger and longer-lasting structure that supports a real understanding of astronomy.

Of the many people who have contributed in some way to this book, I would most like to thank Eric Mazur, whose work in Peer Instruction for physics has inspired science teachers all across the world, and who continues together with his Galileo project to encourage and facilitate the extension of Peer Instruction to chemistry, astronomy and other science disciplines. Thanks to Suzanne Brahmia at Rutgers for being the first to mention Peer Instruction to me. Thanks to Lillian McDermott and the Physics Education Group at University of Washington in Seattle for my early positive experiences with Collaborative Learning in a laboratory setting. The Harvard Science Education Department has provided input and advice; special thanks go to Phil Sadler, Bruce Ward, and Judy Peritz. Himel Ghosh at the Harvard-Smithsonian Center for Astrophysics provided some welcome programming help in reformatting the ConcepTest database. Thanks also to composer and teacher Howie Frazin (Longy) for an early critical reading, and *most* especially to Beth Hufnagel (Anne Arundel Community College) for a careful review, useful suggestions, and insightful comments.

I am delighted to name the many astronomy faculty members and researchers in the field who have contributed to the library of ConcepTests, by submitting questions or ideas, and by reviewing, editing, and testing ConcepTests. The ConcepTest Library for *Peer Instruction for Astronomy* has been and continues to be a community effort, where any instructor is encouraged to use and contribute ConcepTests.

The following colleagues have been the most helpful and encouraging, and their community spirit and generosity have expanded the range and depth of this book. These are Fran Bagenal (U.CO), Bernard Bates (U.Puget Sound), Andreas Berlind (Ohio State), Caroline Cox (U.VA), Steve Danford (UNC, Greensboro), Rodney Dunning (Wake Forest), Peter Edmonds (Harvard-Smithsonian Center for Astrophysics), Kenneth Gayley (U.IA), Peter Garnavich (Notre Dame), Margaret Hanson (U.Cincinnatti), Joe Heafner, Robert Hill (Maranatha Baptist Bible College), Jennifer Hoffman (U.WI), Doug Ingram (Texas Christian), Vinay L. Kashyap (Harvard-Smithsonian CfA), Craig Kletzing (U.IA), Amy Kolan (St. Olaf), George Kraus (College of Southern Maryland), Jodi McCullough (Lisbon High School, Salem, OH), Peter R. McCullough (U.IL, Champaign/Urbana), David McDavid (Limber Observatory), Tony Morgan (Dickinson College), Gerald H. Newsom (Ohio State), Alberto Noriega-Crespo (JPL), Bryan Penprase (Pomona College), Eric Sandquist (San Diego State), Edward G. Schmidt (U.NE, Lincoln), Stephen Schneider (U. Mass), Liliya Williams-Rodriguez (U.MN), Katherine Wu (U.FL), Mike Vaughn (Northeastern), and Maria Womack (St.Cloud).

Many thanks to Alison Reeves and Christian Botting of Prentice Hall, who encouraged me and helped this project become a book. Finally, my deep abiding gratitude above all to my wonderful family, to my loving wife, and faithful friends.

PJG
Cambridge, MA
May, 2002

Chapter One

INTRODUCTION

One of the most challenging aspects of teaching introductory science courses is that many students, expecting to be spoon-fed "the facts," fail to think critically and begin to lose interest. Good teachers know that teaching is a lot more than just dispensing information. These days, information is cheap. In the age of the internet, information leaps out at us at rates exceeding our capacity to absorb or to remain engaged. But the nature of learning has not changed. Understanding and comprehension come only when a student actively turns information around conceptually to view it from different angles. Since this takes time and effort, many students are naturally more comfortable just memorizing phrases and formulas. This is an especially tempting preference in science and math classes, where many great and subtle ideas have indeed been distilled into "laws" and equations. Students' preference for using equations to simply "plug-and-chug" toward an answer is a living legend, long battled by valiant science teachers. We want our students to be able to think and reason, because without this their knowledge is so bounded and fragile that it may never be applied beyond the final exam. One jolting example that illustrates the urgency of teaching students *to think* was provided by a nationwide test of high school students taking part in a National Assessment of Educational Progress (NAEP; Carpenter 1983):

> **Students were asked to *estimate* the answer to 3.04 × 5.3 and choose from the following responses:**
>
> **a. 1.6   b. 16   c. 160   d. 1,600   e. don't know**
>
> **For 17-year-olds, just 40% chose the right answer. When asked to actually calculate the answer, 80% got it right.**

Students often prefer concrete, task-oriented learning to critical thinking and abstract reasoning. It's no wonder, since in many films, TV shows, and classes, scientific *results* are presented as a *fait accompli*, without any indication that real people toiled to puzzle them out in a sometimes serendipitous, often messy adventure of experiment and reason.

> *How can instructors present science as an*
> *evolving endeavor of human curiosity?*

The experimental basis of science is not well grasped by many students, especially those with little previous exposure to the scientific method and how it is applied in astronomy. Many students think that science is just a body of knowledge, a list of facts to be memorized. For some, science seems like an imposing monument built by dowdy and eccentric geniuses laboring alone in ivory towers, basement laboratories, and top-secret government facilities. Instead of experiencing the satisfaction of challenging their own preconceptions to arrive at a deeper understanding, students are often numbed by the blizzard of laws, rules, and facts. The social, collaborative adventure of science is masked.

> *How can we achieve better real-time, adaptive*
> *interaction between instructors and students?*

Instructors, after years of scientific training, can have trouble empathizing with students' challenges, because to the instructor much of the material has become intuitively obvious. Quite understandably, instructors may focus on the topics they perceive to be the most (1) interesting, or (2) daunting, be it from their own experiences years ago, or simply from last year's class. Unfortunately, this year's class may immediately grasp subjects that the instructor belabors, or the class may be overwhelmed by ideas assumed to be obvious. The end result is that students lose interest, become frustrated, or get lost.

Not as subtle is the sheer boredom that can overtake students when they sit in a large lecture hall for an hour or more. During that hour, attention and concentration flags about 10 minutes after they settle into their seats, and may not recover until the last few minutes of class. Student attention spans reliably last only 10–15 minutes. Unfortunately, the most typical method to keep students' attention is stress — the perceived threat that "this might be on the exam." But stress does not enhance students' interest or involvement, and is no remedy for boredom.

Good teachers know their subject well, and communicate their excitement about it. Lecturing is a tried and true method of delivering information, and need not be a passive experience for the *engaged* student. Typically, however, lecturing is about as effective as reading in transmitting information, and somewhat less effective in promoting thinking, or a change of attitudes or beliefs. Many students *postpone* their learning through note-taking. Note-taking focuses on capturing facts, equations, vocabulary, and what might appear on an exam. Good note-takers record lecture highlights and mnemonics for later review. While lecturing is best for teaching facts and principles, discussions and active engagement are better than lectures for promoting higher-level reasoning, positive attitudes, and motivation (Costin 1972; Goldsmid and Wilson 1980). Innumerable educators and theorists have shown that learning is more than memorization. First, the student must be awake and attentive, a challenge in itself that already recommends interludes of activity and discussion. Second, some form of *conflict* should be encountered and overcome that serves to restructure cognition. More than half a century ago, Piaget (1950) proposed that disequilibrium in a student's cognitive structure is what motivates progress from one stage of cognitive reasoning to the next, and he was surely not the first to say so. Conceptual conflict creates curiosity and is essential for discovery learning (Johnson, Johnson, and Smith 1996). This requires active engagement in learning, not simply passive attention.

> **How can we ensure that every student has a personal contribution to the learning process?**

The problem with lectures has now reached the status of a proverb: "Information passes from the notes of the professor to the notes of the student without passing through the minds of either." Why? Lecturers usually assume that all students benefit from the same information, presented at the same pace. This wastes a lot of class time telling students things they could read and understand for themselves. Especially in large classes, students are deterred from asking questions for a variety of reasons, even if the instructor encourages them to do so. Some students are simply shy, and some are being courteous to the lecturer. Other students are insecure about their level of understanding. If they think it's low they don't want to embarrass themselves. If they think it's high, they may not want to seem like a nerd, a "teacher's pet," or a show-off usurping other students' chance to participate.

Standard lecture mode encourages an environment of passivity in the classroom. Keeping students engaged can be a real challenge, no matter how lively and clear your lectures are. But student apathy is doubly destructive. Students who are inactive and unengaged will doze, daydream and consequently retain little. That glassy-eyed, slack-jawed look we know all too well saps our innate energy and enthusiasm for the lecture, which can in turn further degrade students' attention — a potentially vicious cycle.

Especially in courses for non-majors, students may not be strongly motivated by the danger of a low grade.   On the other hand, in a competitive environment, many students have no measure of how well they know the material, or where they stand relative to their peers until exam time.   Students, particularly those who find few role models around them or at the front of the class, may become intimidated. As a result for some, this is their last science class. Rapid personal feedback can overcome many of these problems.

THE ROLE OF PEER INSTRUCTION

As any teacher knows, the real test of learning is to be able to explain what you've learned to others. Regardless of the subject matter, when students are actively involved, they learn more and retain it longer than when they try to absorb knowledge passively. Small-group learning techniques are now in use in science classes all over the country. Students are becoming accustomed to this method in high schools, and will not be surprised to see it in introductory college astronomy. Various forms have been tested and evaluated, and are even now being sculpted by experience in the class and laboratory.   Students who work collaboratively report more satisfaction with their classes (Beckman 1990; Cooper et al. 1990; Johnson, Johnson, and Smith 1991a).   This form of learning has a variety of names and implementations falling under the classification of collaborative learning (CL), including cooperative learning, collective learning, peer teaching, peer learning, Peer Instruction, team learning, study circles, study groups, work groups, and so on.

Group work can take the form of informal learning groups, formal learning groups, or study teams.   The latter two usually are formed for projects that span many classroom or laboratory sessions, with the aim of completing a project. *Peer Instruction for Astronomy* focuses on what instructors can do in class to boost student learning and satisfaction, and so the emphasis here is on informal learning groups.   These are temporary clusters of students, formed as needed, often within a single class session.

These informal groups help gauge students' understanding of the material, allow students to apply what they are learning, and provide a change of pace. Peer Instruction (sometimes called peer teaching) is a form of collaborative learning.   Peer Instruction has been developed and implemented for introductory Physics by Eric Mazur at Harvard University. The improvements in student performance have been widely publicized in Sheila Tobias' book *Revitalizing Undergraduate Science* (Research

Corporation 1992). By encouraging student participation and interaction during lecture, Peer Instruction encourages students to critically think through the arguments being developed, and to discuss and defend their ideas and insights with their neighbors. At any time and in a class of any size, you can implement Peer Instruction. For instance, you may simply ask students to turn to a neighbor for 2 minutes to solve a puzzle or question you've just posed during a lecture.

The goal of this book is to facilitate the implementation of Peer Instruction in introductory astronomy classes by providing ideas, guidelines, and a wealth of examples ready for your use in the classroom. First, I'll outline active learning techniques like Peer Instruction, and the wealth of usage and research behind them. When you first undertake to incorporate Peer Instruction, you will not be alone! I also provide detailed classroom "recipes" for the implementation of Peer Instruction, with specific examples. A broad library of ConcepTests forms the core of the book. ConcepTests are short, conceptual multiple-choice questions for use during class, that serve to gauge comprehension of scientific principles in real-time, to challenge misconceptions, and to foster student engagement through Peer Instruction. Along the way, I provide quick summaries of how to implement Peer Instruction in the classroom for the harried instructor. In addition, I encourage you to participate in a broad range of assessments. I describe options for assessing your students, and for evaluating your own implementations of *Peer Instruction for Astronomy* using input from students or colleagues. I also invite you to join a broad-based collaboration of instructors working to enhance Peer Instruction by building and improving the library of ConcepTests, and by assessing and improving Peer Instruction itself using data from the field— the reported experiences of you and other astronomy instructors.

Chapter Two

PEER INSTRUCTION
FOR ASTRONOMY

The model of a brilliant scientist lecturing imperiously to intimidated students in a sink-or-swim struggle for the shores of knowledge no longer defines good-quality instruction. Appeals for education reform are ever more common. Conscientious instructors focus on such concerns as how inspiring and effective their lectures are, whether the students find them an appealing performer, and whether their delivery is well paced. But from a point of view more suited to learning, they could instead be asking: How inspired are my students? How well connected are they to me, to each other, and to the learning process? How can I tell if they are understanding the material? In a course that makes heavy use of Peer Instruction, some instructors must consider re-framing their role in this way. By using ConcepTests and getting real-time feedback you can directly and effectively experience your own role in the learning process. While this adjustment of viewpoint and shift of strategy may sound stressful, it can actually provide some relief. While you know and probably attempt to communicate that the lion's share of learning is the student's responsibility, the difference is that with Peer Instruction, you really help make it happen.

ACTIVE LEARNING:
ENGAGING THE EGO AND THE MIND

Learning in an active mode is more effective than passive learning. A student who thinks about a concept, phrases it in her own words, and discusses it with others is much more likely to remember and internalize the concept than if she has only heard it. When students talk, listen, read, write, and reflect, the combined power of these learning experiences is greater than their sum. A wide variety of recent research on collaborative and cooperative learning techniques supports this, and is leading to a shift in the old paradigms of the lecture classroom. As described by Mathews, Cooper, Davidson, and Hawkes (1995), the teacher adopts the role of facilitator, coach, or midwife, rather than "sage on the stage." Teaching and learning become experiences *shared* between the teacher and the students. Social interactive skills and the ability to collaborate and work together are boosted. Articulating ideas to their peers illuminates students' own assumptions and modes of thinking. They formulate the concepts in their own vocabulary, and also have an immediate chance to use new terminology, both of which are key to learning. Students' new responsibility for learning and teaching stimulates their intellectual development, and enhances their participation in the process. Most importantly, rather than learning simple facts or even specific abstract concepts, active collaborative learning means that students *learn how to learn*. This is the fundamental goal of any good educational process. After college, most of their own learning and advancement will occur in cooperation with others, and this is modeled only in collaborative learning in the active classroom.

> *There aren't any embarrassing questions,*
> *just embarrassing answers.*
> **(Carl Rowan)**

Almost all students interact with their peers differently than with the instructor. No matter how friendly or approachable the instructor, peer interactions often yield more frank and animated discussions than those with the instructor. The most important questions are "stupid questions"— those whose answers students feel embarrassed not to know. These are the questions that to the student seem the most fundamental, the most likely to unmask their ignorance or unearth broad misconceptions. And while it is never stupid to ask a question, students are much more likely to ask "stupid" questions in a small group of peers than with the instructor and the whole class as audience. Students are motivated to expend more effort before class learning on their own if they know their work is going to be scrutinized by peers. They will also learn the course material in greater depth if they are involved in teaching it to fellow students.

COOPERATION AND COMMUNICATION

Children spend about 34 minutes per day using a computer on average, but when combined with use of other "screen technologies," such as television and video games, children with access spend an average of 5 hours a day in front of a screen. Potential risks associated with excessive use include hampered social development, links to greater depression, and loneliness (Subrahmanyam et al. 2001). If children's social and collaborative experience is waning, the more interaction and cooperation they experience in their education, the better. The college level is for many the last chance to learn those behaviors in an educational setting.

When students graduate and seek employment, they will benefit greatly from cooperative work habits and good communication skills. While some programmers, scientists, or engineers may work mostly alone with their computers, or in their workshops and labs, there are three major reasons for fostering higher-level interactive skills. First, good "people skills" are a prerequisite to any higher-level job involving management and teamwork. Second, given the vast scope of technical knowledge required in the modern technical workplace, collaboration is an absolute necessity. Finally, isolation in the workplace is often alienating and breeds dissatisfaction.

The American Institute of Physics (AIP Education and Employment Statistics Division 1995) recently surveyed employers about their employees with undergraduate physics degrees. When asked what characteristics they valued most in their new hires, the most common responses were the ability to communicate effectively and to work in a team, even above technical skills.

CRITICAL THINKING

Describing their own understanding and contrasting the descriptions of their peers, students feel that their contributions are valued (Meyers and Jones 1993). At the same time, they are called upon to practice and develop higher-order thinking skills of analysis, synthesis and evaluation. The ability to think critically, to make ballpark estimates, and to deal with poorly defined problems are all prerequisites for success and advancement in any professional job, and especially in a technical career. By emphasizing conceptual problem-solving rather than rote memorization and calculation, Peer Instruction addresses these needs.

BENEFITING FROM DIVERSITY
AND ENHANCING IT

Peer-group discussions engender a lot of student-to-student interaction and active participation. In the traditional classroom, when a question is posed by the instructor a single student is called upon to respond, while other students remain passive. By comparison, during peer discussion interludes, a large fraction of the class is engaged in debate and explication. Students get to know one another better, further enhancing participation. Students learn by teaching other students. Rather than hearing only a single prepared explanation from the instructor, the diversity among students and among their groups highlights many ways to approach a problem.

The Census Bureau has made clear that "minorities" will soon be in the majority in many states. University enrollment in the United States is expected to rise about 20% between 1995 and 2015 to 16 million, with minority students comprising 80% of that increase (Carnevale and Fry 2000). Hispanic-American enrollment will increase from 11 to 15%, Asian-Americans from 5 to 8%, while the fraction of African-Americans is expected to remain steady at 13%. Overall, minority student enrollment in undergraduate programs will increase from 29.4 to 37.2%. Diversity in the educational setting is not just ethnic and cultural. Less than half of students enrolled in colleges and universities fall into the traditional 17 to 21 year age range (cited in *The Chronicle of Higher Education Almanac* 1996). Many students are coming back to school, are working part-time jobs, or have family obligations. Other students may lack the self-esteem necessary to compete in all-class discussions of challenging concepts, preferring to let the teacher or the more confident students take the floor.

While many introductory college astronomy courses have mostly non-science majors, some are geared toward science majors or prospective majors. Unfortunately, attrition from science classes is often viewed by instructors as a process of natural selection for ability or motivation. However, at least a third of students switching out of a science, math, or engineering field cite as a primary reason for leaving that their "morale was undermined by competitive culture" (Seymour 1993). Signs are positive for a better representation of women in physics. In 1999, women earned 21% of the physics Bachelors, an increase of 2% from the previous year, and 10% from 1978 (Mulvey and Nicholson 2001). Further progress is important, and achievable. Studies of the role of gender in class participation (e.g., Gardner et al. 1989), particularly in American under-graduate science classes, show that small-group work increases the participation of women in science classes, and should boost their enfranchisement in the field.

For all these reasons, the most effective teaching strategies are increasingly versatile and adaptive, engaging students with the instructor and with each other. These strategies accommodate a diversity of backgrounds, and also take advantage of the focused approach of older, more experienced students. Since Peer Instruction is teacher-guided but implemented by the students themselves, it is well suited to meet these challenges.

KEY COMPONENTS
OF COOPERATIVE LEARNING

Two key elements needed for Peer Instruction to work are positive interdependence and individual accountability. Interdependence means that participants benefit mutually from each other's effort, and that everyone suffers when anyone is left behind. Traditionally, students either work alone, or compete directly with one another. In a well-planned collaborative learning environment, all students contribute to the learning process in the class. They are told at the outset of class and rapidly become aware that their participation is essential for the peer group to achieve and for their class to move forward. Positive interdependence only works in tandem with individual accountability. Both slackers and overachievers are detrimental in a learning environment where each person is responsible for learning and understanding the material at hand. To reinforce these notions, instructors are encouraged to *avoid grading on a curve.*

The research into cooperative learning techniques like Peer Instruction is reassuringly wide and deep, boasting a long history as well (Johnson and Johnson 1993). Some useful references can be found throughout these chapters. These and other references are compiled in the **Readings and Resources** chapter.

UNEARTHING PRIOR KNOWLEDGE

Seasons are caused by the changing Earth-Sun distance. The phases of the Moon are caused by the Earth's shadow. There's a "dark side" of the Moon. Telescopes are used for magnification. The Sun will "explode" at the end of its life. High-mass stars last longer than low-mass stars. Black holes suck in everything around them. He was "light years ahead of his time." These and many other misconceptions have been noted not just anecdotally, but in a variety of well-known larger-scale studies. Sadler's (1992) study and subsequent work by Project STAR highlighted some of these common misconceptions among high school students, while Schneps' famous video "Private Universe" (Schneps 1987) revealed that only 2 of 23 graduating seniors at Harvard were able to explain what causes the seasons. Only 22% of graduating education majors could do so (Atwood and

Atwood 1996). Neil Comins' (e.g., 2001) work on misconceptions provides not only a useful compendium of common astronomical misconceptions, but also a framework for understanding their origin, based on his technique of encouraging students to speculate about, for instance, how many stars are in our Galaxy. Fundamental misconceptions about astronomy form an irregular conceptual lattice to which new information adheres poorly if at all. But when students learn of their own misconceptions, they are motivated to reorder them, crystallizing a stronger foundation for learning.

Zeilik et al. (1998), analyzing introductory astronomy pre- and post-test scores on a multiple choice concept test, found that students had lower initial scores on astronomy concept understanding than physics concept understanding. Sounds grim, except that astronomy concepts proved easier to change. The key then, is teaching effectiveness, along with knowledge of — and a means of addressing — misconceptions. Instructors agree on this. From a long list of elements potentially affecting successful student learning, Comins (2000) asked Astronomy 101 instructors to choose the most important three. The resulting elements named (in decreasing order of votes) were

1. Teaching technique
2. Student prior knowledge/misconceptions
3. Student motivation
4. Student time spent on class
5. Teacher's abilities

But while instructors tend to agree that pre- and misconceptions strongly affect learning, some statistical research does not play along so well. In Zeilik's (1999) results from a large conceptually based astronomy course, students show large gains in conceptual understanding, but no relationship was found between course achievement and completion of previous courses in math or science.

A different view challenges that student misconceptions are a key perspective on student knowledge. Clement, Brown, and Zeitsman (1989) instead focus on the existence of productive resources in students' understanding, noting that "not all preconceptions are misconceptions." When learning about science, students must "construct" new knowledge from previous knowledge. They start with some raw material of past knowledge, experience, and intuition, many of which are valid and applicable in a limited range of situations. As Hammer (2000) points out, for most students who answer that seasons are caused by the changing Earth-Sun distance, they are not likely to be expressing a long-held "misconception." It's more likely that they have not thought deeply before about what causes seasons, and are looking for the first reasonable answer based on their life experience (e.g., *closer is stronger*). *"Closer is stronger"* is a useful resource, but not the proper one to activate here, as you might show by pointing out that when it's summer in the northern hemisphere, it's winter in the south. Another life experience (the noonday sun is strongest), viewed as a resource, could be used to reorder the initial untested response.

Hammer and others encourage teachers to find and use existing student resources. So the best teaching of science is not simply to expose misconceptions and illustrate the proper path to the correct interpretation. It is to build the vocabulary and the conceptual resources of students to have the power and robustness of those used by a practitioner of science. As a scientist, when confronted with a new problem, you bring to it a range of resources for thinking about physical situations. Given a familiar problem, you efficiently apply the right tool for the job. For an unfamiliar problem, you peruse your tool chest, perhaps trying several. You compare the results from each, and work to reconcile conflicts. Using Peer Instruction, you encourage and coach debate in science class to develop these techniques — already natural to you, but much less familiar to most of your students. Well-suited ConcepTests help pit alternative concepts against each other in the crucible of debate, including perhaps some initially plausible preconceptions, or posing some *"extreme gedanken"* — thought experiments at the limits of considered variables. Peer Instruction helps students understand the importance of exploring a variety of perspectives, of critical judgment, and of testing assumptions.

OVERCOMING POTENTIAL BARRIERS
TO PEER INSTRUCTION

Faculty responding to their own drive for excellence, seeking a refreshing departure from standard lecture style, or an evolution in their relationship with students should enjoy experimenting with Peer Instruction in their classes. Implementing Peer Instruction demonstrates both to colleagues and students a commitment to teaching, making Peer Instruction a fresh approach that easily returns the investment. While Peer Instruction is an experiment that can be introduced incrementally if need be, both you and your students will benefit from a more wholehearted embrace. The community of astronomy instructors is eager to work with you: As evidenced by their embrace of the ConcepTest database, they are eager to contribute to and benefit from each others' experience. But the greatest beneficiaries will be your students — from the empowerment that they will feel, and the added comprehension that they will acquire. Instructors may feel a bit daunted by the prospect of implementing Peer Instruction, and adopting any new classroom style requires overcoming some old habits and new anxieties. This is not just resistance to change. Some of the objections that instructors offer are completely valid.

> *There's a lot of material to cover in my course. If I turn the floor over to the students several times each class, how can I expect to cover it all?*

The amount of material that is *heard* in class may indeed go down if several breaks for Peer Instruction are included. But the amount of material that is *learned* should increase, because active learning is a more personal experience. Rather than preparing a monolithic script from which to read throughout the class period, you can begin to focus your preparation on identifying the key concepts. Incorporating Peer Instruction allows you to focus on the Big Picture.

> *Thinking in terms of how much the student is learning as opposed to how much material has been presented is a fundamental and necessary shift in perspective.*
> **(Sutherland and Bonwell 1996)**

Actually, when you use ConcepTests to get a quick read of student understanding, you will sometimes find that students are already up to speed on some topics, which enables you to skip significant parts of some prepared lectures. These are just the parts you *should* skip, since they waste time and put the class to sleep. Leaving this material to the students frees up class time. To make sure they come to class prepared, practitioners of Peer Instruction adopt some simple strategies. Some of these are outlined in Chapter Three, **Recipes for the Classroom,** that is designed to make the transition to Peer Instruction easier. The promise of occasional quick reading quizzes encourages students to complete the assigned reading before class. Homework assignments that enhance class preparedness can also help. A technique that serves double duty for Peer Instruction is to have students do the assigned reading and write up a ConcepTest question of their own with relevant answers. Since some of their results may be used in class, students will be more engaged and interested in the exercise, which leverages your own investment in Peer Instruction. Additionally, you may ask students to prepare ConcepTests based on some topic that is *not* covered in the reading. This should ensure that they have not only read the material, but gone beyond it, especially if you count these student contributions in their overall grade.

For large classes that also subdivide into lab or discussion sections, these provide an excellent arena for further exploration of more difficult concepts. If the instructor is supported by teaching assistants, they can attend classes to follow the ConcepTest discussions. This provides the opportunity for them to note where difficulties lie, and if the instructor does not find sufficient time to elaborate on the topic, it should become a goal for the following discussion section.

> *If I turn the class over to the students, chaos will reign,*
> *and it may become a struggle to regain control.*

Yes, you can expect that class will get a little noisy. Consider this a healthy sign; a third or more of the class is now actively engaged in learning! Most instructors find that it is easy to regain control, because students are eager to hear the correct answer. Also, early on in the course, you should describe the format that you will use and its intent.

Once students become used to the drill, the novelty of Peer Instruction decreases somewhat, so the initial chaos generated by the change of pace and shift to social interaction also diminishes. The group discussions will be time-limited. When students know they have a very limited time for the debate, they typically dig right in. The same time limit should be set to preclude extensive downtime for the groups that finish quickly. A visible clock or even a bell or alarm can serve to accelerate the transition from student- to instructor-led class time, but normally instructors focus attention back to the front of the class whenever the discussions begin to peter out. At that point, either most students have chosen an answer and are satisfied, or everybody is stumped and it's time for you to intervene!

> *Students don't want to be taught by other students.*
> *Somebody is paying for this education, so the students*
> *want to hear from an expert, an authority on the topic,*
> *not from their neighbor.*

Students *will* hear from an expert. What they will not hear is someone reciting out loud what they can read in their textbook. When the instructor's emphasis is on concepts, and when the students are engaged as active participants in their own learning process, students often overcome their initial skepticism with the novel approach. Since many forms of cooperative learning are being used increasingly throughout primary and secondary education, increasing numbers of students will feel comfortable with the method. It's an excellent idea to spend part of the first class describing the method and its rationale, and do allude to the wealth of research and experimentation backing it up. You can even provide some references for students who might be interested in education, or in understanding more about the benefits of cooperative learning. A list of references and readings later in the book should facilitate this.

> ### *Group work lacks rigor and students will take it easy whenever they can.*

The structured small groups used in Peer Instruction should foster both positive interdependence and individual accountability. The research consistently shows that cooperative learning of this type enhances student achievement in nearly every case. In the rare instance where no measurable improvement is seen, these techniques yield similar results to more traditional lecture-based instruction (Davidson 1990).

> ### *Some of my colleagues will be suspicious of this non-traditional classroom style. They'll think I'm unprofessional, or simply trying to avoid lecturing.*

Don't hesitate to let your colleagues know in advance that you are trying a new technique in your classes. Many of them will have heard something about cooperative learning or about Peer Instruction more specifically. Workshops on cooperative learning techniques are conducted all over the country, and the positive experiences of faculty are increasingly discussed at meetings of the American Astronomical Society, the Astronomical Society of the Pacific, the American Physical Society, and the Association of Physics Teachers, to name a few. The rare colleague that resents innovation is likely to be one who prefers that his or her own inertia not be highlighted by the contrast.

From the point of view of the learning institution, the primary litmus tests of a good instructor are that the students learn the material, that they evaluate the instructor positively, and that they are enthusiastic about the course so that enrollment remains high. Since Peer Instruction can help in all these domains, it is worthwhile to emphasize that there may be greater danger in stagnation than in innovation. Teaching leaves more freedom to innovate than is allowed in most professions. You are rarely scrutinized in class by anyone other than your students, so when class has begun, the opportunity for innovation and experiment is as wide as you make it. In the case of Peer Instruction, the innovations are tried and true, so that the case is even easier to make.

Chapter Three

RECIPES FOR THE CLASSROOM

Don't panic.   There is no need to throw away your lecture notes and radically restructure your class for the next semester.   Implementation of Peer Instruction can happen gingerly if you prefer.   Of the many college and university classes using cooperative learning techniques, most employ them between 15 and 40% of the available class time (Cooper 1990, ref. in Millis and Cottell, p. 14).   These techniques supplement, but do not replace, direct instruction and lectures.

DAY ONE

Experienced instructors know the importance of a strong first class to set the course off in the right direction.   In any course, Day One is your chance to set the tone by conveying your natural enthusiasm for the material, describing to students the broad sweep and direction of the learning adventure before them.   Discuss the importance of astronomy and the goals of the course.   List the resources to be used, for example, the textbook, web sites, a lab notebook.   Provide a syllabus for the material to be covered in the course, and cover the usual administrative details.   As you gain experience with Peer Instruction, it's not a bad idea to include some flexibility in the syllabus.   For example, some instructors include one lecture period each for midterm and final exam review.   You may find this "padding" useful if you need to adjust your pace or schedule.

On Day One, just cover the usual important items: course prerequisites, adding and dropping courses, waitlists, your office hours, best methods of contacting you, hours and location of sections if you have them. Lay out the grading scheme.   Make clear your school's and your personal policies on honesty and plagiarism. At many institutions (for example Georgetown University, U. of Maryland, Oberlin College, Wesleyan University), students sign an honor code at the beginning of the course. Students unused to collaborative learning may need to be shown explicitly the boundaries of collaboration.   For work that should *not* be collaborative, say a take-home, open-book exam, some teachers require short signed statements like *"I have neither given nor received any improper aid on this exam."*

On Day One, get to know the students, at least in a broad sense. Who grew up where they could see the stars best? Worst? How many stars do they suppose are visible with the naked eye? These kind of questions can break the ice and make students comfortable, while opening a window to some relevant course material. Offering some kind of background knowledge probe is an excellent idea. Not only does it help you assess the overall class level at the outset, but it also helps students know where they stand. You can even use the individual results to group students by previous knowledge. By comparing results to a similar test at the end of the course, you can later assess the effectiveness of your teaching strategies, and students can more immediately appreciate how much they've learned. One well-tested diagnostic assessment of introductory astronomy students, the **Astronomy Diagnostic Test**, is discussed in Chapter Six on **Assessment** and reproduced in its entirety in Appendix II.

For Peer Instruction, Day One is particularly important. This is when you can set the tone for a relaxed classroom environment where inquiry and participation are encouraged. It is also the best time to make clear that Peer Instruction is not a free-for-all. During the first several classes, lay out clearly what you expect from the students, and how they will be evaluated. Some students may be concerned about being guinea pigs in a novel teaching experiment. Make sure they understand that Peer Instruction is a form of cooperative or collaborative learning, with a long history and proven results. Others may worry that they will spend less time learning and more time teaching fellow students. Reassure them with the arguments already discussed. Students will want to know how they will be graded, and how group grades change the dynamic. Some detailed suggestions for overall class grading are presented in Chapter Six on **Assessment**. To take the mystery out of Peer Instruction, walk them through the outline below, and try an example ConcepTest before the end of class. Students will appreciate having a concrete experience of a new and unfamiliar method, and they will also enjoy a taste of some substantive material during the first class session.

Accept that Day One might feel a little awkward. Especially for beginning instructors, this is common on the first day of any class, but may be more so if you are teaching new material, or using a method that you haven't tried before. It's helpful to appear confident at the outset. Arriving early and chatting with a few of the students can help you relax. What you feel as nervousness often comes across as enthusiasm and energy!

A BRIEF RECIPE

Briefly, lectures are broken into sections covering key points. Start with a more-or-less standard format mini-lecture on one of the fundamental concepts to be covered. This mini-lecture might last about 10 minutes, and is then followed by a ConcepTest, a short multiple-choice question that tests the students' understanding. After one minute, you may ask the students to record or display their individual answers. Recording the initial answers affords the opportunity to track the improvements in understanding that Peer Instruction later builds. You may then ask students to turn to their neighbors to try and convince them of their individual answers. This invariably leads to animated discussions. After another minute or so, the students are asked to reconsider their answer and record it again. You then take a quick poll to decide whether to move on to the next concept, or to continue exposition on the same material. There are a variety of options available to suit your taste; some are sketched below. The process lecture/test/discuss/retest, may repeat several times until the end of the class. Depending on the material, you may thus expect to cover 3–4 key points during a typical one hour lecture period. When you implement Peer Instruction in your classroom, a good plan might be to break your lecture outlines into 3–4 subsections. As an example, a lecture on quasars can be broken up as follows:

1. How nearby active galaxies differ from normal galaxies
2. Evidence for supermassive black holes
3. Quasar distances and luminosities
4. The epoch of quasar formation

Before class, you can choose (or compose) a couple of ConcepTests for each key point you plan to cover. Following your mini-lecture on one such point, the briefest possible use of a ConcepTest might be as follows — simply as a real-time gauge of class comprehension.

ConcepTests for Feedback Only
High Comprehension

1. Mini-lecture
2. Quick-read tally via ConcepTest; yields >90% correct answers
3. Identify and explain the correct answer
4. Move on

After every ConcepTest, you should allow a moment for an explanation, even when the vast majority of students chose the correct answer the first time, without recourse to peer discussions. First, an explanation should be available to those students who did answer incorrectly.   Second, some students will glean the correct answer without true understanding, either from wording, context, or from watching others.

After a ConcepTest, you may instead discover that comprehension is so low that you feel the students should not try to convince each other of the correct answer since so few of them know it.

ConcepTests for Feedback Only
Low Comprehension
1. Mini-lecture
2. Quick-read tally via ConcepTest; yields <20% correct answers
3. Continue mini-lecture, allow for greater detail, review and questions
4. Re-tally with a new ConcepTest to gauge comprehension

The figures of 90% and 20% are of course simply suggestions. Use your own judgment. If the comprehension is intermediate, as is most often the case for well-chosen ConcepTests, then Peer Instruction comes fully into play. Here are some time estimates as a guide:

ConcepTests with Peer Instruction
1. Mini-lecture (10 minutes)
2. Pose ConcepTest (1 minute)
3. Quick-read tally (1 minute); yields 30-80% correct answers
4. Students break into peer groups for discussion (2 minutes)
5. Re-tally after discussion (1 minute)
6. Iterate or move on

If students have already had their first peer group discussion, and a tally shows that a significant but not overwhelming fraction (say half) of the groups found the right answer, then you can ask each group to combine with the nearest group that has chosen a different answer. For a concept this knotty, I suggest allowing about four more minutes of discussion for the new large groups to arrive at a single answer.

For Peer Instruction, the largest and most crucial investment of instructors' time is in choosing good ConcepTests that fall in the middle group, allowing students to teach each other most effectively.   This generates the greatest student engagement, but also relieves you from having to cram material into a full-time lecture, since you now emphasize key concepts over rote learning.   I cover hints for constructing good ConcepTests later on, and a primary goal of this book is to provide many of them, so that the skids are greased for your foray into Peer Instruction.

Now you can see that in the most common situation, covering a key topic should take just about 15 minutes of class time, even allowing for the real-time feedback and the student interaction and discussion that Peer Instruction provides. Although the back-and-forth with students may seem to throw a wrench into the clockwork of a traditional lesson plan, your adaptation will be easier than you might think. Wander around and listen to the discussions and debates going on.   While the students deliberate, you will have some time to think, and an opportunity to evaluate where any confusion might lie and how to address it.

HOW TO GAUGE STUDENT UNDERSTANDING IN REAL-TIME

Counting students' answers rapidly can be difficult in a large classroom. For the first go-round, there are several quick ways to gauge what fraction of students are up to speed with the current course material.

1. **A SHOW OF HANDS**. It's difficult to avoid the herd instinct if you ask students to raise hands sequentially for each possible answer. Rather than risk looking ignorant, many students will try to follow their peers; they will hesitate, hoping that they can take a quick poll to see if the current answer is the right one. Just knowing that they have the option of following the majority may decrease their motivation to think deeply.   One way around this is to have students all simultaneously hold up the number of fingers representing their answer (for answers 1, 2, 3, 4, or 5). Of course, some students may still add or subtract fingers while you're looking to another corner of the room.   "Dipsticking," where students put their heads down on their desks and hold up one hand with the answer, can alleviate some herding there. Unfortunately,

the fraction of correct answers may be hard to gauge by scanning fingers in a large class.

2. **FLASHCARDS**. Flashcards have the advantage that unless they look around very carefully, students cannot gauge others' answers to influence their own. All students can vote simultaneously, and these two benefits remove much of the herd instinct. Flashcards that have printed answers on them may be hard to tally at a glance. One good solution is to use colored flashcards for color-coded answers; rather than offering answers coded A, B, C, D, and E, just label them red, orange, blue, green, and purple.

I have found that flashcards that are white on one side and colored on the other work best, because cards can be raised high, while students looking forward see only white. These can be cheaply and quickly manufactured on the first day of class. As students come into the classroom, they pick up five colored and five white index cards, and glue them together for their own use during the course. (Use glue sticks or spray, and face the lines inward.)

Even simpler flashcards use large bold letters. Students can hold them up in front of their chests to discourage a herd tendency in the class response. To make them more easily identifiable from a distance, large bold patterns can be used. If you (or better yet, students) have access to the web and a color printer, you can use pages that combine all the above properties — large block letters, in color, with white on one side. I have posted these at my web site for downloading and printing as needed:

www.harvard.edu/~pgreen/educ/Flashcards.html

3. **ELECTRONIC CLASSROOM VOTING TECHNOLOGY**. Several useful electronic systems have been developed for instant tallying of student responses, offering class statistics and display as well. These involve small devices for each student that transmit choices to a single computer, which can be tallied, graded, graphed, or projected. For instance, all student answers could be instantly added, and a normalized histogram of their answers projected onto a screen. After peer discussion takes place, the new histogram could be overplotted showing the change.

The most complete system would also allow for identification of individual students, weighting of student answers by self-reported confidence level, grade storage and manipulation. Past answers and improvement rates could be stored for each question, so that the next time the subject is taught, the instructor can evaluate the class and its progress relative to previous classes.

Currently, several electronic interactive classroom communication systems are available. The *Classtalk* system provides for direct wiring of student devices to a single computer running the *Classtalk* software (www.bedu.com/classtalk.html). A wireless alternative is the *Personal Response System* (PRS), developed at the University of Hong Kong's Center for Enhanced Learning and Teaching (www.ust.hk/celt/). PRS uses hand-held, ID-coded transmitters similar to TV remotes, and is distributed in the U.S. by *Educue* (www.educue.com/Home.htm). *Reply* Wireless Response Systems also (www.replysystems.com/) offers several high-end options, as does *Hyper-Interactive Teaching Technologies* (www.h-itt.com). Systems offer different numbers of answer keys (usually 5 or 10), the ability to express the confidence of the answer, the option of anonymous or user-identified responses, and response receipt confirmation. Total costs for these systems span a wide range, from about $1,000 to $30,000 depending on the system, options chosen and class size. The development of less expensive systems using off-the-shelf (e.g., TV) remote control devices would be handy, but registering answers individually with these currently presents a technological or at least a programming challenge.   The ease of classroom setup, the flexibility of the grading and statistics provided, and the compatibility with other software familiar to you are obviously important factors in choosing a system.

If you and your students have individual, live web connections available during class, they can submit answers with a simple Java form, the results of which can be tallied at the front. Such systems are not difficult to develop and host for individual classes, and can be accessed using unique usernames and passwords. The obvious, and overwhelming drawback is the myriad distractions afforded by web access during class!

HOW TO SELECT GROUPS

After the quick read has been taken, if the class answered less than about 80% correct, you will break up the class into peer groups. Generally, this can be very informal, with students turning to those sitting nearest them. There are several more formal ways to group students, described below. As the discussions continue, walk around the classroom, listen in, ask guiding questions, and generally facilitate the discussion. You will rapidly develop a feel for this activity, which may differ quite a bit between your classes. Very often groups will develop their own leadership as students teach each other.

Small groups can be formed in three basic ways: self-assignment, random assignment, or grouping criteria. I recommend self-assignment at least initially, but I also describe a variety of ways to select groups.

SELF-ASSIGNMENT allows students to form groups that are personally compatible. Many students are genuinely interested in learning, and most will at least want to avoid repeated wrong answers in the classroom. This means that most students prefer to collaborate with peers who are at or above their level. Students with the best grasp of the material will most likely wind up with other students at a similar level. In this way, well-matched groups should coalesce naturally.

On the other hand, you may find that self-assigned groups socialize too much, or that students segregate too much into groups with similar cultural and economic backgrounds. However, the comfort provided by self-assigned groups can be a particular asset early in the course when some students are still adapting to the novelty of collaborative learning. Random assignment, or deliberate assignment to groups by criteria, can be phased in later in the course if necessary, after students are more used to Peer Instruction.

RANDOM ASSIGNMENT is most useful in situations where students have a lot to gain from varying group composition, or where specific grouping criteria may be unavailable or difficult to implement. Random assignment can help break up strong social segregation or stratification within the class, better mixing students by gender, background, or academic rank.

Students can draw a group number randomly from a hat passed around the room. They can be assigned groups alphabetically by name — start with A, and add students alphabetically until the desired group size is reached. To regroup differently, you might use, for example, the second letter of the first name. The same thing can be accomplished by birth date — that is, in numerical order, DDMMYY. Some mixing of random and self-assignment is afforded when assigning by birth date — for example, allowing students to form any group of three where all students were born on days beginning with either 0, 1, or 2. You could randomly assign group numbers to birth dates in a table displayed at the front of the class.

Group members can be retained for the entire course, or changed at every lecture according to class needs. But some of these methods may be too time-consuming to be used for each class. Perhaps the most efficient way to randomize is to display a seating chart at the front of the class, so that students are seated next to other group members when it comes time to start a group discussion.

GROUPING CRITERIA allow the instructor to fine-tune student interactions. For instance, students could be assigned to groups by students' prior achievement or preparation, participation, work habits, age, ethnicity, or gender. If you choose to have groups of longer duration (e.g., groups stay together throughout the course), some flexibility will be needed. You may need to reorganize such groups to keep sizes stable, or to improve their functioning. One effective but simple technique to use for short-term grouping is to ask students to "forge a consensus with somebody holding up a different answer."

GROUP SIZE affects students' participation level. In larger groups it is difficult to ensure that every student engages in the process. Smaller groups are less diverse and may not include enough viewpoints to stimulate discussion or to solve the problem (Rau and Heyl 1990). However, since in-class group discussions should last a couple of minutes at most, groups of three or four are ideal. Small group sizes also help ensure that everybody participates. Some teachers find that students often form groups that are larger than requested, so set a firm limit on maximum size.

Any of these methods may make for some chaos in the classroom on the first go-round. You should describe how groups will be formed before you form them. Make it clear that everyone should participate, and that each group must find its own way to handle unproductive behavior. As a way of encouraging full but diplomatic participation, you might even allow students to report on unproductive or domineering partners, which could detract from a portion of their grade allocated to class participation. You can also advertise in advance that after some interludes of Peer Instruction, an occasional pop quiz will be given to students to be completed individually. Some ideas for grading are discussed later in this chapter. But first consider how the experience and success of students with Peer Instruction will improve if they are exposed to a few guidelines for participating in Collaborative Learning groups.

HOW TO BUILD STUDENT COLLABORATIVE SKILLS

Students should be encouraged to discuss why they choose the answers they do, and what their reasoning is. This includes admitting if they have no clue as to the correct answer. In the less intimidating context of Collaborative Learning, many students will naturally be willing to reconsider their own judgments and opinions. But students should also be explicitly asked to listen carefully and openly to comments of each member of their group. Analogous to the social sciences, students will be collecting "qualitative data" — the explanations and the narratives of their peers — whose meaning contributes to and should be analyzed together with each student's own concept of the focus topic.

Group decision-making can easily be dominated by the most aggressive or confident student, by the loudest voice or the most verbose student. Every group member must be given an opportunity to contribute his or her ideas before a final answer is determined.

For groups to be truly cooperative and productive, several elements are key: **(1) positive interdependence, (2) individual accountability,** and **(3) social skills** (Johnson, Johnson, and Holubeck 1990).

1. **Positive interdependence** means a shared goal. This can be implemented by requiring group members to agree on both the answer and the explanation for each problem posed. On occasion, you can ask groups to write down their response and reasoning, and turn it in for a grade, with all members signing. Since any ConcepTest discussion may then turn into a "pop quiz," students are motivated to attend, and to work together to understand the problems. But not all individuals will feel accountable with this method.

2. **Individual accountability** can also be increased simply by reminding students that *anyone* may be called on to explain their group's response. You can announce that on occasion, the pop quiz will be structured so that each student writes down the group's explanation individually, but then you will randomly select whose response is counted for the whole group.

3. **Social skills** are key to group success, but are rarely emphasized in science classes. Asking groups to assign specific roles can be helpful, especially if roles are rotated. Roles appropriate to small discussion groups include a facilitator/mediator, a secretary/foreperson, and a fact-checker/devil's advocate. The facilitator/mediator resolves personal or style conflicts, ensures that everyone participates and has a say, and helps the jury arrive at a verdict. The secretary/foreperson can record both progress and sources of confusion, but above all summarizes and presents the group's choice and explanation. The fact-

checker/devil's advocate is responsible for making sure the group has considered alternatives and resolved conflicts and differing conceptions.

Since good ConcepTests will often present several seemingly reasonable answers to a puzzle, students will encounter some conflict in the social dynamic of their peer group. Collaborative efforts to resolve the conflict require skills that the instructor can enhance simply by pointing them out. Instructors should outline the rules of engagement early in the course before the students begin practicing Peer Instruction. Some instructors may like to post a list of conflict resolution skills during the student discussions. The following is a synopsis of guidelines culled from the experience of many instructors:

1.   **READ.** Come to class prepared to discuss the material.
2.   **RISK.** Be open with your opinions and your questions. Listen to and encourage everyone's ideas so they can take risks too.
3.   **RELAX.** Don't take criticism of your own ideas personally. Change your mind when the evidence shows that you should.
4.   **RESPECT.** Act toward your peers as you would have them act towards you. Be civil, be charitable.
5.   **REASON.** Play the skeptic, but be critical of reasoning, ideas, and data, not of people. Consider the extreme cases to try to understand the principles.
6.   **RESTATE.** Try to paraphrase another's explanation if it is unclear to you. Try to put them together in a way that makes sense. Focus on coming to the best possible answer.

A number of books and web pages worth consulting on skill building for cooperative learning are listed in Chapter Eight, **Readings and Resources**.

HOW TO FACILITATE DISCUSSIONS

Key to Peer Instruction is that the instructor be able to step down from the helm as team captain and onto the sidelines as coach. During Peer Instruction, the instructor no longer scores the goal, but rather cheers on the student teams as they do. This nurtures their appetite for learning, and keeps them from going astray during either their reasoning or their collaborative processes. When peer discussions begin, you can get out of the spotlight and move around. Walk around the classroom. Eavesdrop on students' conversations to learn where the kinks and bottlenecks to understanding are, and chime in when appropriate. Once you give the answer, discussion often stops. But there's no need to keep completely quiet. As a facilitator, you can maximize students' learning by responding directly to their inquiries, but preferably when they have exhausted their own logic and knowledge. You can and should give frequent feedback, but always probe the students' reasoning process, and encourage them to critique it themselves. A more detailed discussion of this role can be found in Wilkerson (1994).

How to provide challenging but still supportive encouragements seems obvious to some instructors, but for many of us it is a challenge. Instructors should raise additional questions that challenge students to think further, without quizzing them or chastising them for shallow thinking. Students don't need to be babied, but if they become embarrassed, they will be less likely to participate. Without being patronizing, you add value to students' experience by acknowledging if their point of view is reasonable, and appreciating an outlook that had never before occurred to you. Here are a few favorite phrases that are constructive in this way:

1. For this situation, have you ever considered or thought about...
2. How did you come to that conclusion?
3. What assumptions are you making to decide that?
4. How is this related to other information we've learned?
5. If what you suggest is true, then how do you explain...
6. That's something I've never thought of before!

Two problems to watch for in groups are conflict and apathy. Impatience, interruptions, aggression, derision, exclusion, and mutiny are all manifestations of conflict and frustration within a group that are generally not hard to recognize. Sometimes conflicts are caused by structural problems. Is the question ambiguous? Ask students to submit written suggestions for improvement that point out the ambiguity, for possible credit. Is the question too difficult given the knowledge that the students have acquired? Ask the students what else they think they need to know, or give a hint towards a more profitable line of reasoning. Is the allotted time too short? Give what extension you can afford. Students need sufficient time to make guesses, float ideas, and assimilate new concepts. Are student levels of understanding too disparate within the group? Address the benefits of diversity in the groups head-on, and offer suggestions on how individual students can profit from it.

Students who feel they fully understand a given topic may on occasion become bored, apathetic, or angry, especially before they are used to Peer Instruction and its benefits. Point out for these more advanced students that teaching is learning, that there will be plenty of opportunities to form other groups on other days and to accumulate grade points. These more advanced students can facilitate comprehension for others. Suggest that they incorporate their knowledge and their observations of common misconceptions into a carefully constructed ConcepTest for extra credit.

For those students whose understanding lags, encourage them to be brave and speak up, to point out what doesn't make sense and try to discover together with other students where the gaps begin. They too can be challenged to go the extra mile beyond acquiring understanding for themselves. Once they feel they understand the topic, they can write and submit ConcepTests illuminating the potential misconceptions. Solicit their written suggestions for revisions to ConcepTests that might have better addressed their confusion. After all, newcomers to town often give better driving directions than do longtime residents!

All students should be made aware that their major responsibility is to work together to do their best. The key to successful conflict resolution is for students to acknowledge the conflict, identify it, and work through it. Encourage students to follow the guidelines of **READ, RISK, RELAX, REASON, RESTATE.**

Chapter Four

CONCEPTESTS

Short, conceptual, multiple-choice questions can be used for two purposes simultaneously: feedback and learning. *Feedback* gives you the ability to quickly gauge student comprehension during class, allowing real-time adaptation of the lecture. *Learning* is facilitated by challenging students to reorder their preconceptions and confront their misconceptions by discussing conceptual puzzles with peers in a collaborative atmosphere.

ACCESSING CONCEPTESTS

One of the most labor-intensive parts of using Peer Instruction is the creation of a large collection of appropriate multiple-choice "puzzlers" of this type. A significant number would be needed simply to cover all the many major topics spanned by most beginning college astronomy courses. However, since class levels vary dramatically both within and between institutions, an even larger collection is advisable. The sample of ConcepTests provided in this chapter is meant to ease an instructor's entry into *Peer Instruction for Astronomy*. This collection contains contributions from instructors across the country, and should be considered a truly collaborative, ongoing community effort of astronomy educators who are interested in progress and innovation in the classroom. The ConcepTests Library remains accessible on the web at

http://hea-www.harvard.edu/~pgreen/PIA.html

Astronomy instructors can both access and contribute to this library. Furthermore, as discussed later, feedback from instructors on the content and scoring of individual ConcepTests will be used to continually adapt and refine the Library in the future, making it a dynamic, accessible tool suitable for direct use in the classroom, but also as a potential *database* for research on and assessment of the technique of Peer Instruction and its results.

CREATING CONCEPTESTS

The ConcepTests provided here and at the web site, though numerous, still provide just a sampling of questions for each topic. You will surely need to write your own as well. What are the characteristics of a good ConcepTest?

A good ConcepTest question will

1. Use scientific principles in a concrete example.
2. Identify and challenge potential misconceptions.
3. Test one concept at a time.
4. Include all, and only, relevant information.
5. Be free of highly technical or inessential jargon.
6. Avoid giving unintentional (e.g., grammatical) clues to the answer.
7. Include most of the reading so that the answers are short and easy to compare.

More generally, if you write negative questions, be careful to underline or capitalize the negative: *"Which one of the following stars is NOT likely to evolve to a black hole?"* If it's feasible, have a colleague or at least a teaching assistant review the ConcepTests you write. Obviously, avoid language that students may find offensive or stereotyping.

Good responses for ConcepTests will

1. Contain a single correct or a single best answer.
2. Be selectable without detailed calculations, or mathematical formulae.
3. Be about equally plausible.
4. Be independent of, and avoid overlapping with, other responses.
5. Be presented in logical or numerical order.

Avoid "All of the above" responses. These are usually the easiest questions, and show that the composer's imagination has dried up. Use "None of the above" advisedly, since while these questions are usually more difficult, they give fewer diagnostics on misconceptions. Avoid absolute language, such as "never" and "always" as a means of making options incorrect. Distractors can be useful. These are incorrect or incomplete response choices that are designed to appear plausible, to be tempting and thought-provoking. When writing responses, you can consider as distractors responses that are correct but do not answer the question posed.

It makes sense to write and use a wide variety of ConcepTest types, to allow for approaches well adapted both to the range and breadth of physical concepts, and to a wide range of learning styles. Mark Schlatter (2001) identified five kinds of ConcepTests that he wrote for his Multivariable Calculus classes: visualization, comparison, translation, theorem-using, and theorem-provoking. These categories are useful to think about when selecting existing, or writing new ConcepTests. The table below, adapted slightly for astronomy, illustrates some general uses for the different types, with several examples specific to the field.

ConcepTest Type	Tests and Develops Comprehension of	Examples in Astronomy
Visualization	3-D thinking, physical configurations.	Size and distance scales, morphology, geometry, optics, orbits, magnetic fields, dots on an expanding balloon as galaxies.
Comparison	Order of magnitude approximations, estimation of relative importance of forces.	Size and distance scales.
Translation	Application of laws and concepts in new, but similar or analogous circumstances. A familiar problem in a new reference frame.	Earthrise from the Moon, seasons on Neptune, parallax of your thumb and of stars, spiral density waves, and sound.
Concept-Using	How and when to apply a physical concept or law.	How would earth's orbit change if the Sun were a solar-mass black hole? Are emission lines from an anti-hydrogen atom different?
Concept-Provoking	Limits of applicability of a physical concept or law and need for new concepts in qualitatively different situations.	Why can't the Sun rotate at the rate of a pulsar? Does the Solar System take part in the cosmological expansion?

Visualization, with its emphasis on observation, is especially relevant for astronomy. The addition of graphics and figures to the ConcepTest Library is therefore an important goal for the future. Animations add even more learning potential for students if they can be projected in class. Interactive web sites with animations and Java applets (HTML-embeddable interactive programs) can be used to illustrate real physical laws and relationships, and allow students to test their understanding and intuition "in the lab."

A method for physics called *Just-in-Time-Teaching* (JiTT) makes heavy use of web-based Java applets (see Christian 1998 and further references below), and is intended primarily as thought-provoking homework assignments submitted just before class. A large compilation of Physlet® Problems can be found at the companion web sites to Douglas C. Giancoli's *Physics: Principles with Applications* and *Physics for Scientists and Engineers,* James S. Walker's *Physics,* and Jerry Wilson/Tony Buffa's *College Physics.* These companion web sites in physics can be accessed at

www.prenhall.com/giancoli/
www.prenhall.com/walkerphysics
www.prenhall.com/wilson

For astronomy, development of a wealth of good applets is even more pressing, since so many of the grand ideas of astronomy cannot be demonstrated at the front of the class the way that physics or chemistry concepts can. The most interesting astronomical phenomena occur in physical conditions unsuitable for the classroom such as the densities of the interstellar medium or of black holes, the pressures that lead to nuclear fusion in the core of the Sun or the core of a supernova, or the floating dance of spiral arms or cluster galaxies. Applets can be used in the classroom for astronomy demonstrations, where students are asked to predict, for instance, *"What will happen to the orbit of the primary star if we double its mass?"* Or *"How will the appearance of this stellar spectrum change if this cool cloud of gas obscures the star?"*

Don't be intimidated by all this blather about what makes a good ConcepTest. Just write! As a conscientious instructor, you surely have a good intuitive sense for most of the goals, and most of the pitfalls. I take the opportunity here to encourage you to submit your class-tested ConcepTests to the Library, posted free for instructors on the web. As the Library grows, it becomes increasingly useful, and allows you to share ideas with other instructors.

A good homework exercise that serves double duty is to have students write and submit their own ConcepTests. This is most effective if students have already participated in at least a dozen peer discussions stimulated by ConcepTests in your class. You should also describe what makes a good ConcepTest prior to the assignment. Student submissions graded for how well they address the goals of ConcepTests will encourage students to think hard about the fundamental concepts, to avoid including throwaway choices, and to reflect on their own learning process and that of their peers.

Student contributions can also prove useful as additions to your own (or the community) Library of available ConcepTests. They may highlight for you misconceptions of which you were not aware. You may wish to assign students to write ConcepTests based on the unit just covered, or the next one up. In the latter case, you may tell students that you will attempt to use one of their ConcepTests in class during the upcoming unit. They will then be more motivated to do the reading thoroughly beforehand. After all, if they provide a ConcepTest good enough for your use in class, they should be guaranteed a few extra points if the responses are graded.

USING CONCEPTESTS

Your awareness of students' ideas and understanding will increase by what students reveal in classroom discussions with peers. Outright guesses, preconceptions, intuitions, and misconceptions can all provide fertile ground for a new ConcepTest, or for revising answers in an existing ConcepTests to better address the relevant issues. During student discussions, roam the room and listen in on their thoughts, and probe the origins of uncertainty or confusion. In some cases, you'll see that a sequence of ConcepTests (e.g., starting with concept-using and followed by concept-provoking types) can be used to help guide students safely deeper into new territory.

Some ConcepTests may seem "tricky" to students. Typically, multiple-choice questions that are tricky may have choices with misleading jargon, humorous or irrelevant distractors, or similar-sounding options that differ by a real technicality not central to the concept being tested. But in ConcepTests, the real trick should be that the choices reflect students' genuine misconceptions about astronomy. On the other hand, a different kind of trickery can keep Peer Instruction fresh, and keep students on their toes. When using ConcepTests in class, you don't have to adhere rigidly to the usual (four choices with one correct answer) format. You can use ConcepTests where more than one answer is valid, without the common choice "more than one of the above." Similarly, you can use a ConcepTest where *none* of the answer choices is correct. By occasionally allowing surprises like these, you will let students know that they need to keep thinking.

As noted in Chapter Three, after every ConcepTest, you should allow a moment for an explanation. For compactness, I have not provided explanations of the correct answers in the ConcepTest Library here. Instructors should be capable of providing them quite easily in most cases. Future enhancements of the ConcepTest Library should, however, include explanations for particularly thorny ConcepTests.

CONCEPTEST FEEDBACK

Feedback from Students

If after a ConcepTest, you call on students themselves for explanations, you will learn more about their thought processes and possible misconceptions. Be careful not to embarrass a student who believes the wrong answer, since that can discourage honesty and future participation as well. Tell students they should feel at ease discussing their train of thought, and critiquing the ConcepTest for content and construction. If the answer is from a peer-group discussion, ask to hear about the group's thought process, so that the individual is likely to feel more comfortable. Direct student feedback about ConcepTests can provide you with great ideas for their refinement, allowing more efficient learning.

If after a ConcepTest, you provide an explanation of the correct answer, then you may also choose to solicit a direct paraphrase of the explanation from your students. You can on occasion have students in-class written paraphrasals count for a grade. Directed paraphrasing is a further tool to access students' comprehension level, and deepens their internalization of the topic. The communication of complex ideas to specific audiences is, of course, an important skill in any profession. As in other facets of Peer Instruction, paraphrasing helps students appreciate the challenges of forming and communicating effective explanations. In the implementation recommended by Angelo and Cross (1993), the instructor identifies a challenging but realistic audience to whom each student's revised or paraphrased explanation should be addressed—for example, next year's class, a group of amateur astronomers, engineering or physics students down the hall. Students should then provide a short written explanation that either replaces or supplements your own, within a limited time frame that you designate, say 3 minutes.

Feedback from Instructors

Every ConcepTest you use in class will rack up a pre- and post-discussion score among students that reveals its appropriateness for this particular class. By recording the pre- and post-Peer Instruction scores that the class achieves for each ConcepTest, you also will be able to learn quickly from experience. You'll be able to tell at a glance if the ConcepTest was appropriate at that point in the course, and also how effective the ConcepTest was at engaging students and promoting cooperative learning.

Class levels can vary significantly between one semester and the next. So the next time you teach the class, you can revise the ConcepTest or its placement accordingly. Or you may find that this year's class generally needs easier ConcepTests. If you choose to keep records of pre- and post-discussion scores, you will be able to accumulate valuable information that can help you choose ConcepTests appropriate to each class you teach. An accumulation of experience with the Library makes it increasingly effective for each instructor who uses it. If the community of astronomy instructors begins to submit statistics on their use of ConcepTests, a rich database will become available that will greatly enhance the appropriate selection and usage of ConcepTests.

Class levels can vary significantly between institutions. If a large enough number of pre- and post-Peer Instruction scores can be accumulated for a given ConcepTest, other instructors will quickly learn what score is most appropriate for their classes, at their particular institution. Instructor feedback to a ConcepTest database can help to increase its effectiveness this way. Check my web page for enhancements that allow your usage and scoring to be recorded, and enticements to do so for the benefit of the community.

Tracking the scoring of ConcepTests is a micro-level of assessment of Peer Instruction. To gauge the effectiveness of your materials and methods overall, to measure your own progress from year to year, and to compare your classes to others', a standard measure such as the Astronomy Diagnostic Test (ADT) is invaluable. See Chapter Six on **Assessment** for other ideas and methods to evaluate and hone your implementation of Peer Instruction for Astronomy.

Chapter Five

A LIBRARY OF CONCEPTESTS

Correct answer choices are denoted by the $ symbol. Electronic versions of the ConcepTests are available in Word and PDF formats on the CD-ROM included with this text.

THE NIGHT SKY

General

The reason stars twinkle is because of motion
  a) on their surface.
  b) of the Earth.
  c) of the Solar System.
$d) of gas in Earth's atmosphere.
  e) relative to the observer.

Which of the following stellar properties can you estimate simply by looking at a star on a clear night?
  a) Distance.
  b) Brightness.
  c) Surface temperature.
  d) Both a and b.
$e) Both b and c.

The Celestial Sphere

According to the heliocentric model, the reason the planets always appear to be near the ecliptic is that
  a) the ecliptic is only 23.5 degrees from the celestial equator.
$b) the planets revolve around the Sun in nearly the same plane.
  c) compared to the stars, the planets are near the Sun.
  d) the planets come much nearer to us than does the Sun.

In the northern hemisphere, the stars rise in the East, set in the West and revolve counter-clockwise around the North celestial pole. In the southern hemisphere the stars rise in the
  a) East, set in the West, and revolve anti-clockwise around the South celestial pole.
  $b) East, set in the West and revolve clockwise around the South celestial pole.

c) West, set in the East and revolve clockwise around the South celestial pole.

d) West, set in the East and revolve anti-clockwise around the South celestial pole.

To see the greatest number of stars possible throughout the period of one year, a person should be located at latitude
   a) 90 degrees.
   b) 45 degrees.
   $c) 0 degrees.
   d) anywhere, since latitude makes no difference.

The "equatorial system" of coordinates
   a) uses the celestial equator as a fundamental reference circle.
   b) uses the vernal equinox as a fundamental reference point.
   c) is "attached" to the celestial sphere.
   d) uses two angles to define a direction in the sky.
   $e) All of the other answers are correct.

The physical basis for the equatorial system of coordinates is
   a) gravity.
   $b) the rotation of the Earth on its axis.
   c) the revolution of the Earth about the Sun.
   d) revolution of the Sun about the center of the Galaxy.

The celestial equator is
   a) the path of the Sun compared to the stars.
   b) the path of the Moon compared to the stars.
   $c) always directly overhead at the Earth's equator.
   d) the average path of planets on a star chart.
   e) always along the horizon for people on Earth's equator.

The ecliptic can be described as the
   a) projection of the Earth's equator onto the celestial sphere.
   b) path of a solar eclipse across the Earth.
   $c) Earth's orbital plane projected onto the celestial sphere.
   d) apparent path of the Moon on the celestial sphere.
   $e) apparent path of the Sun on the celestial sphere.
      (Note: More than one answer possible).

Precession is the
   a) accuracy with which numbers are given in astronomy.
   $b) slow motion of the Earth's rotation axis on the celestial sphere.
   c) apparent backward motion of planets on the celestial sphere.
   e) daily eastward motion of the Sun around the celestial sphere.

If the star Aldebaran rises tonight at 2:00 a.m., when do you expect
it to rise next month?
   a) 11:00 pm.
$b) midnight.
   c) 1:00 am.
   d) 2:00 am.
   e) 3:00 am.

Ecuador is Spanish for equator. It's September 21 and you're in the capitol, Quito.
At noon, how many degrees above the horizon is the Sun?
   a) 0
   b) 30
   c) 45
   d) 60
$e) 90

An object transits (crosses your meridian). For that night, the object has achieved
its highest
   a) declination.
$b) altitude.
   c) azimuth.
   d) airmass.

The zenith distance of Polaris, the "North Star"
   a) is always 90 degrees.
   b) is always 23.5 degrees
   c) is always 0 degrees.
$d) varies with your latitude.

You're stranded on a desert island. You locate the pole star Polaris. It is 17 degrees
above the northern horizon. What is your latitude?
   a) 73 degrees south.
$b) 17 degrees north.
   c) 73 degrees north.
   d) 17 degrees south.

Seasons

In the northern hemisphere, the full Moon transits highest in the sky during
   a) summer.
   b) autumn.
$c) winter.
   d) spring.

The ecliptic makes its smallest angle with the southern horizon during
$a) summer.
  b) autumn.
  c) winter.
  d) spring.

On the first day of winter, the Sun sets
$a) north of West.
  b) directly West.
  c) south of West.
  d) Any of the above, depending upon your location on the Earth.

What causes winter to be cooler than summer?
  a) The Earth is closer to the Sun in summer than in winter.
  b) The daylight period is longer in summer.
  c) The Sun gets higher in the sky in summer.
$d) both b and c.
  e) all of the above.

What is the declination of the Sun on the first day of spring?
$a) 0 degrees.
  b) 45 degrees.
  c) 75 degrees.
  d) The Sun has no declination because it isn't a fixed star.

Northerners have cold days in January because
  a) the Earth is farthest from the Sun in January.
  b) the orbital velocity of the Earth is largest in January.
$c) the Sun is lower in the sky in January.
  d) El Nino is always strong in January.

Suppose you are visiting Australia in August. Which of the following is true? It is
$a) winter and the Sun rises in the northeast.
  b) summer and the Sun rises in the northeast.
  c) winter and the Sun rises in the southeast.
  d) summer and the Sun rises in the southeast.

Imagine a planet whose rotation axis is perpendicular to its orbital plane. How
would you describe its seasons?
  a) shorter than those on Earth.
  b) longer than those on Earth.
$c) constant.
  d) the same as those on Earth.

Time

If the Moon rises at 9pm tonight, tomorrow night it will rise at about
   a) 10:00 pm.
   b) 9:00 pm.
$c) 9:50 pm.
   d) 8:10 pm

If the Earth were in an orbit farther from the Sun than it is now,
   a) the day would be longer.
   b) the day would be shorter.
$c) the year would be longer.
   d) the year would be shorter.
   e) Two of the above are correct.

The sidereal day (a full rotation of the Earth measured relative to distant stars) is 4 minutes SHORTER than a solar day. If the Earth's spin were in the opposite direction then a sidereal day would
   a) not change.
   b) change, but remain shorter than a solar day.
$c) be longer than a solar day.
   d) be the same as a solar day.

The Earth rotates around its axis in a counter-clockwise direction, when viewed looking down on the North Pole. A newly discovered planet revolves in the same sense but spins backwards relative to the Earth. If it revolves in 365.25 days and spins in 23 hours 56 minutes (just like the Earth but backwards spin) which of the following is true? Its day is
$a) less than 23 hours 56 minutes.
   b) between 23 hours 56 minutes and 24 hours.
   c) 24 hours long (just like the Earth).
   d) more 24 hours.

A hypothetical planet whose orbit has a semi-major axis twice that of the Earth's orbit will have a sidereal period of about
   a) 1.0 earth-years.
   b) 0.3 earth-years.
$c) 2.8 earth-years.
   d) 8.0 earth-years.

From the day of the summer solstice to the day of the autumnal equinox, the azimuth of sunrise
$a) shifts toward the East.
   b) shifts toward the West.
   c) shifts toward the South.

   d) shifts toward the North.
   e) remains constant.

From the day of the vernal equinox to the day of the summer solstice, the azimuth of sunset
   a) shifts toward the east.
   b) shifts toward the west.
   c) shifts toward the south.
 $d) shifts toward the north.
   e) remains constant.

A sundial yields
   a) universal time.
   b) local sidereal time.
   c) local mean solar time.
 $d) local apparent solar time.
   e) eastern standard time.

The difference between mean solar time and apparent solar time
   a) varies to plus or minus one hour throughout the year.
   b) is called daylight savings time.
 $c) is called the equation of time.
   d) remains constant throughout the year.

Eclipses

The phase of the Moon at a lunar eclipse
 $a) is always full.
   b) is always new.
   c) is always waning crescent.
   d) is always waxing gibbous.
   e) may be anything depending on the geometry.

The phase of the Moon at a solar eclipse
   a) is always full.
 $b) is always new.
   c) is always waning crescent.
   d) is always waxing gibbous.
   e) may be anything depending on the geometry.

MEASURES AND METHODS

Measures and Units

The expression "order of magnitude" corresponds to
$a) one factor of ten.
  b) one factor of two.
  c) a factor of 2.5
  d) two factors of ten.

You are deciding which computer to buy, based on how long it takes to start running your favorite computer game. Which computer starts your game program the most quickly?
  a) Game starts in 1 centisecond.
  b) Game starts in 1 kilosecond.
$c) Game starts in 1 microsecond.
  d) Game starts in 1 millisecond.
  e) Game starts in 1 second.

Multiplying together two numbers $N \times 10^x$ and $M \times 10^y$ in scientific notation yields the following, again in scientific notation
  a) $(N \times x) \times 10^{M \times y}$

  b) $(N + M) \times 10^{(x+y)}$

  c) $(N \times M) \times 10^{x \times y}$

$d) $(N \times M) \times 10^{x+y}$

What do you get when you multiply 10^3 by 10^3 ?
  a) 0
  b) $1/10^3$
  c) 10^9
$d) 10^6
  e) 2×10^3

At the start of the 21st century, the number of human beings alive on Earth was approximately 6,000,000,000 or
  a) 6 to the 9th power.
  b) 6 to the 9th power of ten.
  c) 6 times ten to the 9th power.
  d) 6,000 million.
$e) both c and d.

The density of an object is its
  a) length divided by its width.
  b) mass divided by its length and width.
  $c) mass divided by its volume.
  d) mass multiplied by its volume.
  e) volume divided by its mass.

A gram of lead has a greater _____ than a gram of feathers.
  a) mass.
  $b) density.
  c) weight.
  d) volume.

Say you need to multiply together measurements of differing precision, some with more significant figures than others. For example, the product of 7.00×6.1 is properly written as
  a) 42.70
  $b) 42.7
  c) 43.
  d) 43.0

Say you want to add measurements of differing precision, some with larger uncertainties than others, such as the sum 4.371 (+/- 0.001) cm + 302.5cm (+/- 0.1). This sum is correctly written as
  a) 306.001 cm.
  b) 306.872 cm.
  $c) 306.9 cm.
  d) 307 cm.

You are parked on an asteroid watching spaceship traffic. You measure the velocity of a spaceship moving away from you to be +50 km/sec. A different spaceship approaches you at the same speed. In the same coordinate system, what velocity do you measure?
  a) +50 km/sec.
  $b) –50 km/sec.
  c) the velocities cancel to 0km/sec.
  d) +100 km/sec.

The Angstrom is the most suitable unit for describing a measurement of
  a) the distance between the Earth and the Sun.
  b) any tiny angle.
  $c) the wavelength of visible light.
  d) the speed of light.

Distance

Sound travels at a speed of 300 meters per second. In analogy to the light-year, what does 1 sound-minute equal?
   a) The time sound takes to travel 300 meters.
   b) The time delay of a sound heard 300 meters away.
   $c) The distance traveled by sound in 1 minute.
   d) The speed of sound 1 minute later.

The star Betelgeuse is about 500 light years away from us in the constellation Orion. If this star underwent a supernova explosion right now, approximately how long would it be until we found out about it?
   a) almost immediately.
   b) 8 minutes.
   c) 10 years.
   $d) 500 years.
   e) 500 light years.

If the speed of light were half what it is now, then a light year would
   a) take half as long to traverse at light speed.
   $b) take the same amount of time to traverse at light speed.
   c) last twice as many months.
   d) last half as many months.

How far from the Earth is the nearest star?
   $a) 1 astronomical unit.
   b) 1 light year.
   c) 1 parsec.
   d) 10 light years.
   e) 10 parsecs.

The trigonometric sine of a small angle between two points on a sphere is approximately equal to the linear separation between those points divided by
   a) the distance to the points.
   b) the radius of the sphere.
   c) the circumference of the sphere.
   d) π
   $e) a and b above.

The distances of nearer stars may be measured by observing their apparent motion as
   $a) the Earth orbits around the Sun.
   b) the Earth rotates on its axis.
   c) the Sun orbits around the center of the Galaxy.
   d) the planets cross their path.

The most important reason for measuring the parallax of a star is to help us find the stars'
    a) direction of motion.
    b) proper motion.
    $c) intrinsic brightness (absolute magnitude).
    d) radial velocity.

If the parallax of a star is measured to be 0.1 seconds of arc, its distance is
    a) 10 astronomical units.
    $b) 10 parsecs.
    c) 1 parsec.
    d) 0.1 parsec.
    e) 0.1 astronomical units.

Two identical stars, one 5 light years from Earth, and a second 50 light years from Earth are discovered. How much fainter does the farther star appear to be?
    a) square root of 10.
    b) 10.
    $c) 100.
    d) 1,000.
    e) the farther star does not appear fainter, since it is identical.

Suppose that you have observed an asteroid over a month. You observe that it is now 25 times fainter than it was on the first day. Estimate how much farther away is it from you now than it was on the first day:
    a) twice as far.
    $b) five times farther.
    c) 12.5 times farther.
    d) 25 times farther.
    e) 625 times farther.

Star A is twice as far away as star B, but they have the same luminosity. This means star B has _____ times the flux of star A.
    a) 1.
    b) 1/2.
    c) 1/4.
    d) 2.
    $e) 4.

You photograph a region of the night sky in March, in September, and again the following March. The two March photos look the same, but the September photo shows three stars in different locations. The star whose position shifts the most must be
    a) farthest away.
    $b) closest.

c) receding from Earth most rapidly.

d) just passing through the Galaxy.

What would you need to observe to find out the distance to a star that was too far away for trigonometric parallax?

a) apparent brightness and intrinsic brightness (luminosity).

b) apparent brightness and temperature.

d) intrinsic brightness and temperature.

$e) apparent brightness, spectral type, and luminosity.

From Earth, planet A subtends an angle of 5 arcseconds, and planet B subtends an angle of 10 arcseconds. If the radius of planet A equals the radius of planet B,

a) planet A is twice as big as planet B.

$b) planet A is twice as far as planet B.

c) planet A is half as far as planet B.

d) planet A and planet B are the same distance.

e) planet A is five times as far as planet B.

For two stars of the same apparent brightness, the star closer to the Sun will generally have

a) a higher flux.

b) a hotter temperature.

$c) a lower luminosity.

e) identical physical properties.

If two intrinsically identical stars are at different distances from the Earth, the more distant star will have a

a) bluer color.

b) higher luminosity.

c) lower luminosity.

$d) lower apparent flux.

Magnitude

Using the magnitude system of astronomy, how would the brightness of an 8th-magnitude star compare to a 7th-magnitude star?

a) 10 times brighter than the 7th-magnitude star.

b) 10 times dimmer than the 7th-magnitude star.

c) 2.5 times brighter than the 7th-magnitude star.

$d) 2.5 times dimmer than the 7th-magnitude star.

Four stars are all members of the same star cluster. Their absolute magnitudes are 10, 7, -1, -10. Which star appears brightest from the Earth?

a) the first.

$b) the fourth.

c) the second.
d) the third.

The reason astronomers use the concept of the absolute magnitude is to allow stars to be compared directly removing the effects of differing
$a) distance.
b) mass.
c) temperature.
d) radius.

The formula m–M = 5 logd –5 would not be true were it not for
a) the fact that distance in parsecs is the reciprocal of the parallax in arcseconds.
b) Newton's law of gravitation.
$c) the inverse square law or radiation.
d) the definition of Vega as zero on the magnitude scale.

Astronomical Methods

Images (or photographs) can be used to measure what properties of a star?
a) location on the sky.
b) distance.
c) brightness.
d) proper motion.
$e) all of the above.

Astronomers determine the "color" of a star by calculating the
$a) ratio of the fluxes as measured with two different filters.
b) difference between the fluxes as measured with two different filters.
c) ratio of the temperature to the radius.
d) ratio of the absolute and the apparent brightnesses.
e) difference between the absolute and the apparent brightnesses.

The color of a star is a number representing the
a) ratio of absolute luminosities at 2 wavelengths.
b) ratio of apparent luminosities (flux) at 2 wavelengths.
c) ratio of the star's area to its distance squared.
$d) Two choices are correct.
e) None of the choices is correct.

TELESCOPES

You observe a star through a 10-inch telescope on Earth. Your identical twin observes the same star from 3 times farther away. How big a telescope does your twin need to make the star appear as bright as it does to you?
  a) 5 inches.
  b) 10 inches.
  $c) 30 inches.
  d) 90 inches.

A telecope's sensitivity (light gathering power) is determined primarily by its
  a) objective diameter.
  $b) objective area.
  c) magnification.
  d) objective focal length.
  e) spatial resolution.

The diameter of the objective of telescope 1 is D. The diameter of the objective of telescope 2 is $2D$. The light gathering power of telescope 2 is ___ times the light gathering power of telescope 1.
  a) 1.
  b) 2.
  c) 3.
  $d) 4.

A telescope's spatial resolution is determined primarily by its
  $a) objective diameter.
  b) objective area.
  c) magnification.
  d) objective focal length.
  e) spatial resolution.

One disadvantage of a refracting telescope compared with a reflector of the same diameter is that the
  a) refractor bends the light more.
  $b) refractor focuses light of different wavelengths at different locations.
  c) refractor has a lower magnification.
  d) refractor has poor spatial resolution.

Which of the following are problems with large refracting telescopes?
  a) weight.
  b) chromatic aberration.
  c) sagging of the objective.
  d) both a and c.
  $e) all of the above.

One Keck telescope on Mauna Kea has a diameter of 10m, compared to 5m for the Palomar telescope. The light gathering power of the Keck is larger by a factor of
   a) 2.
$b) 4.
   c) 15.
   d) 50.

One of the Gemini telescopes has 4 times the light gathering power of the 4-m telescope on Kitt Peak. How many times farther into space can a Gemini telescope detect a faint object?
   a) That distance does not depend on light gathering power.
$b) 2 times
   c) 4 times
   d) 8 times
   e) 16 times

The reason astronomers use telescopes in space is because the Earth's atmosphere
   a) absorbs or scatters some of the light at all wavelengths.
   b) absorbs all of the light at some wavelengths.
   c) distorts images of astronomical objects.
$d) All of the above.

The main reason astronomers use large telescopes is that large telescopes
   a) have a bigger field of view.
   b) have a larger magnification.
$c) gather more light.
   d) None of the above.

Which of the following has never been seen since the invention of the telescope, and probably does not require a telescope to be detected?
   a) an individual star in another galaxy.
   b) a change in the dark spot on Neptune.
   c) a comet striking a planet.
$d) a supernova exploding in our Galaxy.

Light is refracted when passing through a lens because
   a) light is absorbed by the surface of the lens and then re-radiated in a different direction.
$b) light moves at different speeds in the glass than it does in air.
   c) the wavelength changes when the light strikes the glass which causes a bending of the wave.
   d) internal reflections caused by atoms in the lens deflect the light rays.

You are looking at a wall clock through a simple telescope objective lens. Which number will be at the bottom of the image formed in the focal plane of the lens?
   $a) 12
   b) 3
   c) 6
   d) 9

Telescopes have motors (clock drives) to track the stars across the sky because the
   a) stars have proper motions.
   b) Earth moves in its orbit.
   $c) Earth spins on its axis.
   d) Sun moves in its orbit.

Temperature differences between the primary mirror of a telescope and the surrounding air cause
   a) color distortions due to heat.
   b) image blurring due to temperature-related focus changes.
   c) image blurring due to buckling of the mirror surface.
   $d) image blurring due to air turbulence.

Which of the following combinations of focal lengths for the objective/eyepiece would have the LOWEST magnifying power?
   a) long/long.
   b) long/short.
   $c) short/long.
   d) short/short.

The larger of two radio telescopes will collect
   a) less energy, and have greater resolving power.
   b) less energy, and have less resolving power.
   c) more energy, and have less resolving power.
   $d) more energy, and have greater resolving power.
   e) the same energy, and have the same resolving power.

Which of the following cannot be determined using radar?
   a) size.
   $b) mass.
   c) roughness.
   d) distance.

HISTORY

The old geocentric view of the Universe held that the Earth was surrounded by a celestial sphere that held the stars and

a) never moved.
$b) rotated from East to West each day.
c) rotated from West to East each day.
d) took one year to rotate around the Earth.

Explanations of planetary motion were more difficult in ancient times than the
motions of the Sun and moon because the
$a) planets reverse their motion sometimes (retrograde motion).
b) Sun and moon always go in retrograde motion.
c) planets are much fainter than the Sun and moon.
d) planets move much faster than the Sun and moon.

Which two fundamental misconceptions made Ptolemy's geocentric model very
complicated and prevented it from adequately describing the movements of bodies
in the Solar System?
I)   The Sun is at the center of the Universe.
II) All heavenly bodies move in combinations of perfect circles.
III) The Earth is at the center of the Universe.
IV) The stars never move.
a) I and IV.
b) II only.
c) III only.
$d) II and III.

Epicycles are needed in the geocentric model to explain why
$a) planets seem to vary in brightness and speed over the year.
b) some planets are brighter than others.
c) the Sun rises and sets.
d) all planets cannot necessarily be seen on any given night.

Why are the phases of Venus at odds with the geocentric epicycle picture?
a) Venus would not be expected to significantly change size.
b) Venus would not appear "new" as the Moon does.
$c) Venus would not appear "full" as the Moon does.

Galileo's discovery of the moons of Jupiter was shocking because it showed those
who believed the Earth to be at the center of the Universe that
a) a planet could have more than one moon.
b) Earth was unusual to have only one moon.
c) gravity could hold a moon in orbit.
d) Jupiter does not move.
$e) moons could move around a moving planet.

Galileo's observations of sunspots discredited Aristotle's teachings because they
showed that

$a) change in the heavens without uniform circular motion.
  b) planets move on epicycles.
  c) the Earth's orbit is elliptical.
  d) planets orbit around the Sun.
  e) the Sun isn't perfectly spherical.

The revolution epitomized by Copernicus that still dominates scientific thought could be described as having moved our view of mankind's place in the Universe from
  a) the edge towards the Big Bang.
  b) the center to the edge.
$c) a special place to an average place.
  d) an average place to a special place.

Newton explained Kepler's laws
  a) as the "music of the spheres."
  b) by putting Kepler's complicated formulad into words that could be easily understood.
  c) by giving them a sound religious footing.
$d) by showing mathematically that the planets moved according to a few universal principles of motion and gravity.

In the scientific method, observations (data) are most useful for
$a) testing predictions made by theories.
  b) making predictions.
  c) proving theories.
  d) generating catalogs.

The best test of a scientific hypothesis is how
  a) well it explains all known observations.
  b) well it agrees with known theories.
  c) simply it explains all known observations.
$d) well it predicts new observations.
  e) easily it is transcribed into mathematical notation.

Faraday performed crucial experiments, and Maxwell contributed a complete mathematical theory for
$a) the wave description of electromagnetic phenomena.
  b) the wave/particle duality.
  c) wave phenomena in nature.
  d) the transformation of electromagnetic charge into radiation.

GENERAL MOTION/FORCES

Orbital Motion

Kepler's 3d law (that the period squared is proportional to the semi-major axis cubed) does NOT apply to the motion of
   a) a satellite around the Earth.
   b) a comet around the Sun.
   c) one star about another in a binary star system.
   d) one galaxy about another.
   $e) All of the above apply.

A description for the relationship between the period of revolution P and the distance from the center R of a point on a record on a turntable would be

   a) P^2 is proportional to R^3 .

   b) P^2 is proportional to R.

   c) P is proportional to R.

   $d) P does not depend on R.

A spacecraft in one stable orbit moves to another stable orbit that has twice the semi-major axis as the first. The period of the second orbit is ____ times that of the first orbit.
   a) 0.5
   b) 2
   c) cube root of 64
   $d) square root of 8

Since angular momentum is conserved, the rotational speed of a collapsing gas cloud
   a) depends on its mass.
   $b) increases.
   c) decreases.
   d) is independent of its initial rotation.

Since angular momentum is conserved, the orbital speed of a planet at perihelion as compared to aphelion
   $a) is larger.
   b) is smaller.
   c) is the same.
   d) approaches infinity.

A planet at perihelion has a value of X for its angular momentum. When the planet is at aphelion, its angular momentum will be
   a)   less than X

b) greater than X.
$c) equal to X.
d) cannot tell without further information.

If the Sun were suddenly replaced by a solar-mass black hole, the Earth would
$a) remain in the same orbit.
b) move into a smaller orbit.
c) move off the current orbit in a straight line.
d) be pulled into the black hole.

An artificial satellite passes near Jupiter and gains some orbital energy in a
"slingshot effect." What is the effect on Jupiter?
a) cooling
b) none
$c) The effects are minimal because of the small satellite mass.
d) faster rotation
e) Jupiter will revolve about the Sun faster.

A bowling ball and a paper bag are pushed out of the airlock of a tourist spaceship.
The spaceship is coasting with its engines off, halfway along on a weeklong cruise
to the Moon. After ten minutes,
$a) both objects are still moving together alongside the spaceship.
b) the bowling ball is still moving alongside the spaceship, but.
  the paper bag has been left far behind.
c) both objects are far behind the spaceship.
d) the bowling ball is falling back towards Earth, while the paper.
  bag is still coasting alongside the spaceship.

For some reason, you wake up alone in space, twirling a ball on a string, far from
any planet, satellite, or massive object. The string breaks. Assuming you would
care under such circumstances, what path does the ball take?
a) It moves directly away from you in a straight line from the point where the
string broke.
$b) It moves away from you sideways in a straight line from the point where the
string broke.
c) It continues to move in the circle because space is frictionless.
d) It moves in a paraboloid arc, like a baseball on the Earth.

Forces and Acceleration

Which situation(s) does NOT describe an acceleration?
$a) a car traveling with constant speed on a straight road.
b) a car traveling with constant speed around a bend.
c) a planet traveling in its orbit around the Sun.

d) a car decreasing speed on a straight road.

e) an electron traveling around a nucleus.

A piece of toast held at rest above the kitchen floor has

$a) potential energy only.

b) kinetic energy only.

c) both potential and kinetic energy.

d) neither, only thermal energy.

Two identical spacecraft are to be accelerated by rockets. The first rocket fires with a force four times as great as the second. The acceleration of the first rocket is _____ as large as the acceleration of the second.

a) ¼

b) ½

c) 2 times

$d) 4 times

e) None of the above.

Two Toyotas collide and come to a dead stop. How much more kinetic energy is lost in this collision than in one between two identical cars each moving at half the velocity?

a) 2 times larger

b) 2 times smaller

$c) 4 times larger

d) 4 times smaller

e) Unchanged, since the potential energy, not the kinetic energy, depends on velocity.

Electromagnetism is

$a) sometimes a repulsive force and sometimes an attractive force.

b) always a repulsive force.

c) always an attractive force.

d) none of the above.

The electromagnetic force does not dominate interactions between celestial bodies because

a) it is too weak.

b) it acts only over short distances.

c) it only affects magnetic materials.

$d) positive charges cancel the effects of negative charges.

Of the four fundamental forces of nature, the one that is inherently the weakest (with the smallest coupling constant) is the

a) strong nuclear force.

b) weak nuclear force.

c) electromagnetic force.

$d) gravitational force.

Which of the four fundamental forces of nature has the largest effect on objects at astronomical distances from each other?
   a) strong nuclear force
   b) weak nuclear force
   c) electromagnetic force
$d) gravitational force

The force that holds molecules together is the
   a) strong nuclear force.
$b) electromagnetic force.
   c) centrifugal force.
   d) gravitational force.

For the strong nuclear force to hold together two protons, they must be
   a) moving slowly.
   b) moving rapidly.
$c) close together.
   d) of opposite charge.

If the distance between two protons is increased by a factor of 10, the (gravitational/electromagnetic) force between them is decreased by a factor of
   a) 10/100.
   b) 100/10000.
   c) 10000/10000.
$d) 100/100.

You are an astronaut taking a space walk to fix your spacecraft with a hammer. Your lifeline breaks and the jets on your back pack are out of fuel. To return safely to your spacecraft (without the help of someone else), you should
   a) throw the hammer at the spaceship to get someone's attention.
$b) throw the hammer away from the spaceship.
   c) use a swimming motion with your arms.
   d) kiss your ship good bye.

If the strong nuclear force were weaker than the electromagnetic force at the scale of an atomic nucleus then fusion reactions would
$a) not be possible.
   b) happen all the time.
   c) not change, since this is already true.
   d) be caused by the electromagnetic force instead.

Gravity

Gravity is
  a) sometimes a repulsive force and sometimes an attractive force.
  b) always a repulsive force.
$c) always an attractive force.
  d) none of the above.

The force of gravity is responsible for
  a) holding the planets near the Sun.
  b) the tides.
  c) holding the Moon near the Earth.
$d) All of the above.

A person is standing on the roof of a one-story building. The ground floor of the building has to support
  a) no weight.
  b) less weight than the roof.
  c) the same amount of weight as the roof.
$d) more weight than the roof.

The reason that visitors to the South pole do not fall off the planet is that
  a) atmospheric pressure is larger at the South pole.
  b) gravity points down and so do the visitors.
$c) gravity does not come from off the planet, it comes from the center of the Earth.
  d) centrifugal force caused by the planet's rotation keeps them there.

When you are on the top floor of a building, your weight is _____ when you are on the ground floor.
  a) much greater than
  b) very slightly greater than
  c) equal to
$d) very slightly less than
  e) much less than

The *Apollo* astronauts repeated one form of a classic experiment by simultaneously dropping a feather and a hammer on the Moon. The feather and hammer reached the ground at the same time because
  a) gravity on the Moon is less than on Earth.
  b) the force of gravity is larger on the feather than on the hammer.
$c) the acceleration of each object is the same.
  d) the astronauts showed Galileo's experiment to be false.

The escape velocity from the Moon is less than that from the Earth because of the Moon's
   a) lower density.
   $b) smaller mass.
   c) smaller radius.
   d) higher temperature.
   e) distance from the Earth.

If the size of the Earth were to double (with the mass unchanged), the gravitational attraction of the Sun would
   a) double.
   b) decrease by two times.
   c) increase by four times.
   d) decrease by four times.
   $e) not change.

If the radius of the Earth were to double, with no change in its mass, a person's weight would
   a) be unchanged.
   b) increase by factor of 4.
   $c) decrease by a factor of 4.
   d) double.
   e) be cut in half.

Antimatter is identical to matter except that its electrical charge is opposite. Suppose that the Moon suddenly changed to antimatter. How might the Moon's orbit change?
   a) Instead of being attracted to the Earth, the Moon would be repelled and move away.
   b) Since matter and antimatter attract, the Moon would move closer to the Earth.
   $c) Since the Moon is neutral, the orbit would not change.
   d) The Moon and perhaps the Earth would annihilate in a gigantic explosion.

Use your knowledge of the escape velocity equation to answer the following. Planet X is twice as massive as the Earth and four times as large as the Earth. Which of the following is true? The escape velocity of planet X is
   a) greater than Earth's escape velocity.
   b) the same as Earth.
   $c) less than Earth's escape velocity.

If woman A approaches man B at a distance C, the gravitational force between them
   a) exceeds their weight.
   $b) quadruples as C halves.
   c) is proportional to the sum A+B.

d) is proportional to the distance C squared.

Suppose that you are in a spaceship traveling past two different planets. You pass within 1 AU of a blue planet that has twice the mass of the Earth, and within 1 AU of a red planet that has the same mass as the Earth. From which planet will you feel the greater gravitational pull?
$a) the blue planet.
  b) the red planet.
  c) the gravitational pull from each planet will be equal.

If one could magically turn off gravity from the Sun, the Earth would
$a) travel in a nearly straight line along its present velocity perpendicular to a line connecting Earth and Sun.
  b) leave the Solar System along a line connecting the Earth and Sun.
  c) spiral outward from the Solar System.
  d) collide with the Moon.

Compared to your mass here on Earth, your mass out in the space between the stars would be
  a) zero.
  b) negligibly small.
  c) much much greater.
$d) the same.
  e) the question cannot be answered from the information given.

The Moon's tidal forces cause the Earth's rotation to slow and the Moon to
$a) decrease its orbital speed and increase its distance.
  b) decrease its orbital speed and decrease its distance.
  c) slow its rotation.
  d) do nothing (no effect).

At different times during its trip, a space probe passes two moons (X and Y) that are identical in size. Though the probe passes both at the same distance, ground control finds that the probe's trajectory is deflected more by moon X than moon Y. Planet X must
$a) be denser than planet Y.
  b) be rotating more quickly than planet Y.
  c) be less massive than planet Y.
  d) has less of a magnetic field than planet Y.

SCALES OF SIZE, DISTANCE, MASS AND POWER

Arrange the following items according to size, from the smallest to the largest.
$a) atom, planet, Sun, galaxy, cluster of galaxies.

b) proton, galaxy, open cluster, cluster of galaxies.
c) proton, star, galaxy, Solar System.
d) Sun, Solar System, cluster of galaxies, globular star cluster.

The establishment of a reliable cosmic distance scale is a "bootstrap process" because
a) distance steps are all calibrated independently.
$b) each distance step calibrates the next step.
c) scientists build from past work.
d) the Hubble Constant calibrates all the steps.

The largest identifiable structures in the Universe are
a) clusters with hundreds of galaxies.
$b) filaments with galaxy clusters along them.
c) galaxies like our own.
d) giant stars.

The cosmological principle enables astronomers to generalize from what they observe to the properties of the Universe as a whole. The principle states that any and all observers, everywhere in space, should see, on average, the same picture of the Universe as us on scales comparable to
a) the Solar System.
b) the galaxy.
$c) superclusters of galaxies.
d) atoms and subatomic particles.

The distance between the Sun and its nearest star is smaller than the distance from the Milky Way Galaxy to the next nearest large galaxy Andromeda by a factor of about
a) a hundred.
b) a thousand.
$c) a million.
d) a billion.

Comparing the Moon's orbit around the Earth to the Sun's orbit around the Galactic center is like comparing the width of a human hair (about one micron or millionth of a meter) to
a) a human.
$b) the distance between Boston and Washington, D.C.
c) the Moon's orbit around the Sun.
d) the size of the Galaxy.

If the Sun were the size of a tennis ball (6.35 centimeters) then the Earth and Jupiter would orbit at distances of about
a) 7 and 35 centimeters, respectively.

$b) 7 and 35 meters, respectively.
  c) 7 and 15 meters, respectively.
  d) 0.7 and 3.5 kilometers, respectively.

How many planet Earths could fit inside the Sun, which has a radius 100 times larger than the Earth's?
  a) A few hundred.
  b) A few thousand.
$c) About a million.
  d) A billion.

About what angular resolution is required to resolve the image of two stars separated by 2 AU in a binary system, if the binary is only 1 parsec (about 200,000 AU) from Earth?
  a) 10 degrees.
  b) 10 arcmin.
$c) 10 arcsec.
  d) 10 milliarcsec.

THE ELEMENTS

The force that keeps electrons in orbit around the nucleus of an atom is due to
  a) the intense gravity of the dense nucleus.
$b) the attraction of opposite electrical charges.
  c) the magnetic field of the nucleus.
  d) its orbital velocity.

Another name for an isolated proton might be a
$a) hydrogen ion.
  b) hydrogen isotope.
  c) helium ion.
  d) neutron.

The structure of the Bohr model of the atom most resembles the structure of
$a) the Solar System.
  b) the Milky Way Galaxy.
  c) the Universe.
  d) a baseball.

When the Sun and planets were created, there was no element Poodlonium. Labradorium atoms decay to form Poodlonium with a half-life of 800 million years. If you find a rock that is one quarter Labradorium, and three quarters Poodlonium, how old is the rock?
  a)   400 million years.

b) 800 million years.
$c) 1.6 billion years
d) 3.2 billion years

If you could add a proton to an atom to create a new stable, isolated atom, you would have created
$a) a different element with a positive charge.
b) an isotope of the original element.
c) a fission reaction.
d) a neutron and a positron.

If an isotope of a particular element has too many or too few neutrons in its nucleus, it is
a) called a molecule.
b) called an ion.
c) no different than any other stable isotope.
$d) unstable against radioactive decay.

A rock's original composition included 4% of a radioactive element. Now we measure only 1%. About how many half-lives of that element have passed since the rock was formed?
a) 3
b) 96
c) 4
$d) 2

The stable element Zorox is only formed by decay of the radioactive element Linoleum decays. You analyze a rock to find that it has 3 times as much Zorox as Linoleum. The age of the rock, expressed in half-lives of Linoleum is
a) 1.
b) 1/2.
$c) 2.
d) 3.

Chemical elements heavier than iron are created primarily by the buildup of
a) colliding massive subatomic particles in the Big Bang.
$b) neutrons in atomic nuclei during supernova explosions.
c) proton decay reactions in empty space.
d) stable nuclear fusion in the cores of massive stars.

Most of the heavy elements (everything except hydrogen and helium) in the Earth were produced
$a) by stars that burned out before the Solar System formed.
b) in chemical reactions in the primitive oceans and atmosphere.
c) in nuclear reactions in the Sun.

d) in the hot, dense, early Universe.

When the Universe is twice its current age, the fraction of atoms in the Universe
that are hydrogen
   a) will be close to zero.
   b) will be close to one.
   c) should remain approximately constant.
   $d) will continue decreasing.

RADIATION AND THE ELECTROMAGNETIC SPECTRUM

General Properties of Light

A charge at rest can create
   a) electromagnetic radiation.
   b) only electric and magnetic forces, no radiation.
   $c) only electric forces.

A charge moving uniformly in a straight line can create
   a) electromagnetic radiation.
   $b) only electric and magnetic forces, no radiation.
   c) only electric forces.
   d) only radiation, but no electric or magnetic forces.

Electromagnetic radiation will be created by any charged particle that
   $a) accelerates (changes speed or direction).
   b) moves in a straight line at constant speed.
   c) remains at rest.
   d) is subjected to a gravitational field.

Visible light is electromagnetic radiation in the sense that it is
   a) composed of electrons and protons.
   b) composed only of electrons.
   $c) created only by and interacts only with electric charges.
   d) generated by radioactivity.

The wave model of light is better than the particle model for explaining
   a) shadows.
   b) reflections.
   $c) constructive interference.
   d) the inverse square intensity drop with distance.

Which of the following characteristics of light are associated with the particle
nature of light?

a) diffraction, interference, and photoelectric effect
b) Doppler effect and formation of spectral lines
c) Doppler effect, diffraction and interference
$d) formation of spectral lines and photoelectric effect
e) photoelectric effect and Doppler effect

Comparing water waves and light waves,
$a) both spread out after passing through a narrow opening.
b) only light waves can reflect off barriers.
c) only light waves display Doppler shifts.
d) only light waves interfere with each other.
e) only water waves interfere with each other.

Light waves differ fundamentally from either water waves or sound waves because they
a) can have various wavelengths.
c) travel from place to place instantaneously.
d) have an unchanging amplitude.
$e) can travel in a vacuum.

If you were going to design a pair of glasses for seeing animals at night, you would want them to convert
a) infrared photons to X-ray photons.
b) X-ray photons to optical photons.
$c) infrared photons to optical photons.
d) optical photons to UV photons.

The observed speed of light is affected by
a) the velocity of the source of the light.
b) the velocity of the observer.
c) the reference frame of the observer.
$d) none of the other answers are correct.

Which of the following are believed to be absolute concepts that are the same in any reference frame?
a) time and distance
$b) the speed of light and the laws of physics
c) mass and acceleration
d) time, distance, and the laws of physics

Which type of electromagnetic radiation has a wavelength adjacent to, but longer than ultraviolet light?
a) radio
b) infrared
$c) visible

d) X-ray
e) gamma ray

Compared to visible light, radio waves have
   a) higher energy and longer wavelength.
  $b) lower energy and longer wavelength.
   c) lower energy and shorter wavelength.
   d) higher energy and shorter wavelength.

Compared to optical photons
   a) radio photons have a longer wavelength.
   b) X-ray photons have a larger frequency.
   c) infrared photons have a smaller energy.
  $d) All of the above.
   e) None of the above.

The two main "windows" in the Earth's atmosphere that astronomers can use are in the visual and radio bandpasses. This is mostly because other wavelengths suffer from excessive
   a) interference.
   b) reflection.
  $c) absorption.
   d) refraction.

Spectra can be used to measure what properties of a star?
   a) Radial velocity.
   b) Chemical composition.
   c) Surface temperature.
  $d) All of the above.
   e) None of the above.

Continuous Radiation, Emission and Absorption

When something is what we usually call "red hot," it is hotter than something that is
   a) blue hot.
   b) white hot.
  $c) neither of these.
   d) both of these.

A hot, glowing, opaque solid, liquid or gas emits which type of spectrum?
  $a) continuous.
   b) emission lines.
   c) absorption lines.

A hot, glowing, opaque solid surrounded by a cool gas will show
  a) continuous emission.
  b) emission lines.
  $c) absorption lines.
  d) both emission lines and absorption lines.

Assume that a star behaves as a blackbody. If the surface temperature of that star doubles, then the wavelength at the maximum intensity will be _____ times the original wavelength.
  $a) 0.5
  b) 1
  c) 2
  d) 4

Suppose two observers look at the spectrum of a cloud of gas in a laboratory; the first reports seeing emission lines and the second reports absorption lines. How can this best be explained?
  a) The first observer sees the gas against a hot background.
  $b) The second observer sees the gas against a hot background.
  c) One observer is moving rapidly relative to the other.
  d) The atoms in the gas are forming molecules.

A star with a continuous spectrum shines through a cool interstellar cloud composed primarily of hydrogen. The cloud is falling inward toward the star (and away from Earth). Which best describes the spectrum seen by an Earthbound observer?
  a) blueshifted hydrogen emission lines
  b) blueshifted hydrogen absorption lines
  c) redshifted hydrogen emission lines
  $d) redshifted hydrogen absorption lines
  e) a redshifted hydrogen continuum

Suppose the atoms in a receding gas cloud have two energy levels separated by an energy corresponding to 4863 angstroms. The observer will see a spectrum with absorption at a wavelength
  a) less than 4863 angstroms.
  b) equal to 4863 angstroms.
  $c) greater than 4863 angstroms.
  d) 4863 divided by the velocity.

Now consider a cool cloud of gas between a star and observer to be moving away from the source of continuous radiation (and towards the observer). The atoms in the gas have two energy levels separated by an energy corresponding to 3000

angstroms. If the spectral line is observed to be at 2700 angstroms, what is the relative velocity of the cloud with respect to the light source?
- a) 5700 km/s
- b) 300 km/s
- $c) 30,000 km/s
- d) 0 (no motion)
- e) the speed of light divided by 300 km/s

You are gradually heating a lump of rock in an oven to an extremely high temperature. As it heats up, the lump emits nearly perfect theoretical blackbody radiation, meaning that it
- a) is brightest when hottest.
- b) is bluer when hotter.
- $c) is both.
- d) is neither.

Doppler Effect

A star is moving away from the Earth at 200 km/s. What can you say about the wavelength of H-alpha absorption that you would detect in the spectrum of the star?
- a) It would be less than 656.3 nm.
- b) It would be 656.3 nm.
- $c) It would be more than 656.3 nm.
- d) There is insufficient information to decide.

For a sound source passing by at constant speed, the sound
- a) gets higher and higher in pitch as the source approaches and lower and lower as the source recedes.
- $b) are of constant higher pitch as the source approaches and of constant lower pitch as the source recedes.
- c) Both a and b above.

What causes light from a star to be Doppler-shifted?
- a) the distance between us and the star
- b) the gas and dust between us and the star
- $c) the speed of the star toward or away from us
- d) temperature differences between us and the star

We can detect the velocity of a star through the Doppler effect by
- a) measuring the shift in distance of the star.
- b) taking photographs six months apart.
- c) applying the inverse square law of brightness.
- $d) measuring the shift in wavelength of a spectral line.

For a sound source passing by at constant speed, the sound
    a) gets higher and higher in pitch as the source approaches and lower and lower as the source recedes.
    $b) is of constant higher pitch as the source approaches and of constant lower pitch as the source recedes.
    c) is of constant lower pitch as the source approaches and of constant higher pitch as the source recedes.
    d) maintains the same pitch except at the moment when the source passes you.

Absorption lines due to the same atomic transitions in the interstellar medium are sometimes seen at several slightly different wavelengths in stellar spectra. This is evidence that interstellar gas is composed of
    a) one huge flowing cloud.
    b) one huge expanding or possibly contracting cloud.
    $c) several separate clouds moving at different speeds.
    d) many clouds in turbulent motion.

Two stars have the same mass and spectral class. Star A is rotating faster than Star B. The width of the spectral lines of star A appear to be _____ than the spectral lines of star B.
    a) narrower
    $b) wider
    c) stronger
    d) the same

Spectral Lines and Energy Levels

The uniqueness of the spectral line pattern of any element is caused by
    a) the density of the gas in the stellar atmosphere.
    $b) the energy-level structure of the atom.
    c) the temperature of the stellar atmosphere.
    d) the Doppler shift.

Atoms of different elements have unique spectral lines because each element
    a) has atoms of a unique color.
    b) has a unique set of neutrons.
    $c) has a unique set of electron orbits.
    d) has unique photons.
    e) none of the above; spectral lines are not unique to each type of atom.

In the Sun, the transition from level 4 to level 2 of hydrogen produces photons with a wavelength of 486.1 nm. In a star twice as hot as the Sun, this transition would produce photons with

a) half that wavelength.
$b) the same wavelength.
c) twice that wavelength.
d) four times that wavelength.

Atoms have particular associated spectral lines because
$a) electrons have only certain allowed orbits.
b) light consists of waves.
c) light waves can show the Doppler effect.
d) photons have only certain allowed orbits.
e) speed of light in a vacuum is a constant.

If a photon of wavelength less than 912 Angstroms interacts with a hydrogen atom, what happens?
a) The atom recombines.
$b) The atom ionizes.
c) An electron jumps up one energy level.
d) An electron jumps down one energy level.

The number of electrons lost by an atom in a gas (that is, its ionization state) depends primarily on the
a) velocity of the gas.
b) level of the ground state.
$c) temperature of the gas.
d) size of the gas cloud.
e) energy required to strip away all the atom's electrons.

An atom has energy levels with energies of 1, 3, 9 and 10. Energy level 10 represents the outermost bound orbit of the atom. Assuming this atom has a single electron in the 3d energy level, which of the following photon energies can the electron absorb?
a) 1
b) 2
c) 6
d) 8
$e) Two of the above answers.

Atom X has energy levels of 1 and 10. Atom Y has energy levels of 1 and 12. An electron in each atom moves from the upper energy level to the lower energy level, emitting a single photon in the process. Which of the emitted photons has a longer wavelength?
$a) the photon from atom X
b) the photon from atom Y
c) the wavelength is independent of the photon energy
d) cannot tell without more information

THE EARTH

The Earth bulges at the equator compared to the poles (6378 versus 6335 km). What's a possible explanation?
- a) The Earth is squashing under its own weight.
- b) Tidal stretching from the Moon's pull.
- $c) The rapid rotation of the Earth.
- d) Tectonic plate motion stops at the equator.

The Earth bulges at the equator compared to the poles. If a seal weighs 100 lbs at the equator, what do you estimate it will it weigh at the pole?
- a) 37 lb.
- b) 99 lb.
- c) 100 lb.
- $d) 101 lb.
- e) 137 lb.

The main reason Earth's atmosphere stays warm is heat from
- a) volcanoes.
- $b) solar radiation.
- c) the greenhouse effect.
- d) continental drift.

The greenhouse effect is caused mostly by
- a) the aurora borealis.
- b) pesticides.
- c) depletion of the ozone layer.
- $d) carbon dioxide in the atmosphere.

The Earth's core is hot due to
- a) solar heating.
- b) large scale meteoroid bombardment.
- c) radioactivity.
- d) heat left over from the Earth's formation.
- $e) both c and d

The Earth's magnetic dynamo is caused by a combination of convection in its molten core and
- a) the Earth's orbit around the Sun.
- $b) the Earth's rotation.
- c) lunar tidal action.
- d) the aurora borealis.

Particles from the Sun interact with the Earth's magnetic fields to produce
- $a) the aurora borealis.

b)   dangerous ultraviolet radiation.
c)   hydrogen and helium.
d)   protons and electrons.

Over the last 10 – 100 million years, the largest-scale changes on the surface of the Earth were caused by
a)   glaciation.
b)   erosion by wind and water.
c)   man-made pollutants.
$d)   continental drift.
e)   comets and asteroids.

Which of the following statements help explain why summer days tend to be hotter than winter days?
      I.   The day is longer.
      II. The Earth's surface receives more light per unit area.
      III. The light is coming directly from the Sun.
      IV. There is no snow to cool the Earth.
a) I.
b) III.
$c) I, II.
d) I, IV.
e) II, III.

What would the days and seasons be like if the Earth still rotated at the same speed, but the Earth's axis were tilted nearly 80 degrees on its side instead of 23 degrees?
a) Both days and seasons would be half a year long.
b) The days would be the same length, but the seasons would be half a year long.
c) There can be no seasons in this situation, but days would always be 24 hours long.
$d) The days and seasons would the same lengths as ours, but the seasons would have more extreme temperature changes.

The Earth is thought to have formed from
$a)   accretion of small rock-like bodies.
b)   the collapse of a single gas cloud.
c)   gas ejected from the Sun in a massive solar flare.
d)   a disk of gas orbiting the Sun.

The fact that the Earth has undergone differentiation suggests that the Earth's interior
a) has always been solid.
b) must not have a core.
c) causes the continents to move.

$d) was once liquid.

Earthquakes are an important tool for scientists because their vibrations
 a) influence the Earth's orbit about the Sun.
 $b) allow the study of the Earth's internal structure.
 c) inform us of the mass of the Earth.
 d) are used to predict the future of plate tectonics.

Heating of the Earth's surface by the Sun is an example of transfer of energy by
 a) conduction.
 b) convection.
 $c) radiation.
 d) pollution.

The reason that the Earth's surface has so few meteor craters compared to other nearby bodies in the Solar System is that
 a) the Moon blocked almost all of the meteors that otherwise would have hit the Earth.
 b) the atmosphere causes all objects entering from space to burn up before they hit the ground.
 $c) the Earth's surface has been modified by various forces which cover or remove traces of the craters.
 d) all meteorite impacts break the crust and release lava from the mantle to fill in the hole.

Which of the following is evidence that the Earth's interior is not rigid?
 I. Plate tectonics
 II. The liquid oceans
 III. The greenhouse effect
 IV. The magnetic field
 a) I, II.
 $b) I, IV.
 c) II, III.
 d) II, IV.
 e) III, IV.

Over the course of time, what happens to the plates that cover Earth's surface?
 a) Water erosion wears away at the edges of the plates at the same rate at which new rock is added.
 b) Plates just get added onto at the midocean rises, and the surface gets more crumpled as plates continue to run into each other.
 $c) Parts of plates sink into the interior and melt, while other parts are created anew.
 d) Where new rock isn't being added to them, the plates just move relative to each other along fault lines.

The Earth's surface absorbs energy from sunlight and then radiates about as much energy back into space. The best proof you can offer for this fact is that the average surface temperature on Earth is
   a) warming slowly.
 $b) not warming rapidly.
   c) half light and half dark.
   d) warmer at the core than on the surface.

THE EARTH AND MOON

A full moon
$a) can never rise at midnight.
  b) can never transit at midnight.
  c) always rises due East.
  d) always sets due West.

Suppose tonight is new moon. You are on the side of the Moon facing the Earth. What phase Earth do you see?
   a) You can't see the Earth because it is eclipsed by the Sun.
   b) new Earth
   c) first quarter Earth
 $d) full Earth
   e) third quarter Earth

As seen from the Moon, how often does the Sun rise?
   a) never
   b) about every 24 hours
   c) about once per week
 $d) about once per month
   e) about once per year

As seen from the Moon, how often does the Earth set?
 $a) never
   b) about every 24 hours
   c) about once per week
   d) about once per month
   e) about once per year

Assume that the Sun rises at 6:00 a.m. At what time does the Third Quarter Moon rise?
   a) 9:00 a.m.
   b) 12:00 noon
   c) 9:00 p.m.

$d) 12:00 midnight

When the Moon is in its last quarter, when does it rise and set?
   a) sunrise, sunset
   b) sunset, sunrise
   c) noon, midnight
$d) midnight, noon

Why don't lunar eclipses happen at every full moon?
$a) The Moon's orbit doesn't always take it through the Earth's shadow.
   b) The Moon's orbit sometimes takes it far enough away that the Earth's shadow has disappeared.
   c) The Moon's orbit is not circular.
   d) The Moon cannot be full AND pass through the Earth's shadow.

An observer on Earth sees a total lunar eclipse. If someone else is standing on the side of the Moon facing the Earth at the same time, they would the
   a) Earth's night side, not eclipsing the Sun.
   b) Earth's day side, not eclipsing the Sun.
   c) Earth partially eclipsing the Sun.
$d) Earth completely eclipsing the Sun.

During an annular solar eclipse, the Moon's orbital velocity is
$a) slower than during a total solar eclipse.
   b) the same as during a total solar eclipse.
   c) faster than during a total solar eclipse.
   d) the same as the Earth's orbital velocity around the Sun.

An observer on Earth sees a total solar eclipse. If someone else is standing on the side of the Moon facing the Earth at the same time, what would they see? The Earth's
   a) night side.
   b) day side, which doesn't change in brightness.
   c) day side, which becomes completely dark.
$d) day side, with a dark spot that moves across it.

Under which circumstance might there be a lunar eclipse every month? If the
   a) Moon were smaller compared to the Earth.
$b) Moon's orbit was in the ecliptic.
   c) Earth's orbit were in the ecliptic.
   d) Earth's axis was not inclined to the ecliptic.

For a solar eclipse to occur, the Moon must be
   a) high in the sky.
   b) near first or last quarter.

$c) near new.
  d) near full.
  e) over another country.

What change in the Moon might enable us to see its entire surface from Earth?
  a) If the Moon were always full.
  b) If its axis of rotation were tipped slightly.
$c) If it rotated at a different speed.
  d) If its orbit were in the ecliptic.

In which case(s) can the Earth, Sun, and Moon be in a straight line?
  a) full Moon only
  b) first quarter and third (or last) quarter
  c) new Moon only
$d) both full and new Moon
  e) This can happen at any phase of the Moon.

If we see a first quarter moon today, what will people on the other side of the Earth see later today?
  a) new Moon.
$b) first quarter Moon.
  c) full Moon.
  d) third quarter Moon.
  e) Any of the above, depending upon the time of night.

Earth doesn't experience an eclipse of the Sun every month because
  a) sometimes the Moon is too far away.
  b) the Moon always keeps its same side toward the Earth.
$c) the Moon's orbit is not in the same plane as the Earth's orbit.
  d) you have to be in the right place to see a solar eclipse.

A friend exclaims that a few months ago she saw the full moon overhead at noon. Is this possible?
  a) Yes, because a full Moon is caused when the Moon is fully lit by the Sun, which can only happen at noon.
  b) Yes, but only if she was approximately 45 degrees from zero latitude.
$c) No, the full Moon is never overhead at noon.

THE MOON

There is very little atmosphere on the Moon because
  a) dry rocks on the Moon absorbed its own atmosphere.
  b) it was blown away by meteor bombardment.
$c) its low mass and high temperature allowed most gases to escape.

   d) the gravitational tidal forces from the Earth stripped it away.

The Moon
   a) always points the same face toward the Sun.
   b) does not rotate.
   c) rotates at the same rate as the Earth — once per day.
$d) rotates on its axis with the same period as its revolution about the Earth

Which theory of lunar origin lost support when the Apollo landing results showed similar abundances of isotopes of oxygen on both the Earth and Moon?
   a) the fission theory
$b) the capture theory
   c) formation together from the same gas cloud

Place the statements concerning lunar formation below in chronological order from the time of formation:
    I.   coalesced from orbiting debris
    II. cooling of interior
    III. mare formed
    IV. surface melting by heavy bombardment
$a) I, IV, III, II
   b) III, II, IV, I
   c) I, II, III, IV
   d) IV, III, II, I
   e) II, III, IV, I

The lunar mare were formed primarily by
   a) melting and solidification followed by impacts.
   b) volcanism.
$c) impact with later volcanism.
   d) volcanism with later impact.

Why are some large crater walls sharp and steep, while others are more rounded?
   a) Different volcanoes make different craters.
$b) age differences
   c) size differences among the impact bodies
   d) composition differences among the impact bodies
   e) seismic activity on the Moon

During a single night, the Moon
   a) moves from West to East across the sky.
$b) moves from East to West across the sky.
   c) appears fixed in the sky above a given location on Earth.
   d) appears fixed in the sky relative to the constellations.

If the Moon was in its 3d quarter last Saturday, what phase will it be this Saturday?
  a) waning crescent
  b) waxing gibbous
  c) at or very near first quarter
$d) at or very near new
  e) Any of the above, depending upon other factors.

If you observe the Moon rising in the East as the Sun is setting in the West, then you know that the phase of the Moon must be
  a) new.
  b) first quarter.
$c) full.
  d) third (or last) quarter.
  e) Any of the above, depending on other factors

It is full moon and it is the night of winter solstice. The Moon must be
  $a) north of the celestial equator.
   b) on the celestial equator.
   c) south of the celestial equator.
   d) Not enough information is given to answer this question.

Why can we never see an absolutely new moon? It is
  a) always on the other side of the Earth.
  $b) always too close to the Sun.
  c) never up during the day.
  d) always on the horizon.
  e) always covered up by the Sun.

When two continental plates collide, the likely result is that
  $a) folded mountains are created.
   b) a hot spot is produced.
   c) very young rocks emerge from below the crust.
   d) a crater is formed.

Your latest discovery, Planet Eris, never had volcanoes. The processes of erosion on Eris have unequal but constant rates in different surface regions. A reasonable way to estimate which region has the least erosion is simply by choosing the region with
  a) the fewest craters.
  b) the largest craters.
  $c) the most craters.
  d) maria.

The fact that there are fewer craters on the lunar maria than on the lunar highlands indicates that the

a) highlands are younger than the maria.
b) craters formed as a result of volcanic eruptions.
$c) maria formed more recently.
d) maria are of volcanic origin.

The Moon at night appears brighter in the winter because
a) the nights are longer.
$b) the declination of the full moon is higher in winter.
c) the cold air makes the Moon appear stronger.
d) the interaction between the tides and the cold air intensifies moonlight.

THE PLANETS

You visit a planet that rotates on its axis in 2 hours, while it revolves closely around its sun every 25 hours. You construct for the inhabitants an excellent calendar with a leap year that occurs once every
a) year.
$b) 2 years.
c) 12 years.
d) 25 years.

Order the following by increasing radius.
   I. Sun
   II. Jupiter's orbit around Sun
   III. Earth's orbit around Sun
   IV. Sun's orbit around the Galaxy
a) III, II, I, IV
b) I, II, III, IV
$c) I, III, II, IV
d) I, II, IV, III

Mnemonic: My Very Educated Mother Just Showed Us Nine Planets. Which planet(s) can never be seen on the meridian at midnight?
$a) Mercury
$b) Venus
c) Mars
d) Jupiter
e) Saturn
(*Note:* Mark all possible correct answers.)

The surface of Venus is mostly low density rock, while the average density of Venus is similar to that of Earth. The interior of Venus is therefore
a) rapidly rotating.
b) composed of low density material.

$c) denser than the average density of Earth.
 d) homogeneous.

How much more solar energy does Venus receive than the Earth, due to the fact that Venus is 0.72 times as far from the Sun?
 a) 0.72 times as much
 b) $1/0.72 = 1.4$ times as much
 c) $0.72^2 = 0.52$ times as much
$d) $(1/0.72)^2 = 1.92$ times as much
 e) 0.28 times as much.

From Earth, Mercury is difficult to see mostly because it
 a) has low albedo.
 b) is very dense.
 c) is very small.
$d) always appears near the Sun.

Mercury's atmosphere is likely to be weak or non-existent because of its
 a) small mass.
 b) slow rotation.
 c) high surface temperature.
 d) high density.
$e) (a) and (c).

Volcanoes on Mars have become much larger than on Earth mostly because Mars lacks
 a) a thick atmosphere.
 b) flowing water.
$c) plate tectonics.
 d) a large moon.

Satellites of Mars probably formed
 a) out of a merger of telecommunications giants.
 b) out of the same material from which Mars formed.
$c) in the asteroid belt, and were later captured.
 d) in the Oort Cloud.
 e) as a result of cometary bombardment.

Synchrotron radiation is produced
$a) by electrons moving rapidly in a magnetic field.
 b) by electrons traveling with a velocity determined by the temperature of the gas in which they are found.
 c) by a "blackbody."
 d) by microwave ovens.

The atmosphere of Jupiter has few heavy elements due to
  a) evaporation.
  b) the lack of heavy elements in the outer Solar System.
  c) the vacuum-like power of the Great Red Spot.
$d) gravitational settling toward the planet's center.

Jupiter's chemical composition is closest to that of
$a) the Sun.
  b) Earth.
  c) Mars.
  d) the Moon.
  e) Venus.
The shape of Jupiter's magnetic field is flattened because of the solar wind and the
  a) low temperatures in the outer Solar System.
  a) planet's strong gravity.
$c) planet's rotation.
  d) many moons.

Io's volcanic activity is driven mostly by heat from
  a) radioactivity.
$b) the Jupiter-Io tidal force.
  c) Jupiter's magnetic fields.
  d) the solar wind.

Synchrotron radiation from Saturn implies that the planet has
  a) powerful electrical storms.
  b) an unseen companion.
$c) a magnetic field.
  d) excess thermal radiation.

The sizes of particles comprising Saturn's rings are studied by analyzing
  a) shadows cast by the rings.
$b) how light is scattered by the ring particles.
  c) excess radiation emitted by the rings.
  d) excess radiation emitted by the planet that is blocked by the rings.

Uranus is peculiar because its
  a) radius is so large.
  b) orbit is highly eccentric.
  c) orbit is strongly inclined to the ecliptic plane.
$d) axis of rotation is in the ecliptic plane.

Seasons on Uranus are
  a) as long as Uranus' year.
  b) non-existent.

$d) one fourth of Uranus' year, or about 20 Earth years long.
  e) four times Uranus' year, or about 320 Earth years long.

Uranus is surrounded by a cloud of
  a) asteroids.
$b) hydrogen.
  c) helium.
  d) water molecules.

Which of the following is correct?
  a) The rotation axes of planets are perpendicular to their orbital plane (the ecliptic).
$b) No planet's orbital inclination is more than 20 degrees from the ecliptic.
  c) Planets orbits are circular.
  d) Planets always rotate and revolve in the same direction.

Which of the following are correct?
 $a) The Earth is closest to the Sun in January.
  b) The Earth is farthest from the Sun in January.
  c) The Earth is always at the same distance from the Sun.
  d) It's cold in winter because the Earth is farther from the Sun.

Which of the following are correct?
  a) Mars is essentially in the same orbit as the Earth.
  b) There are canals filled with liquid water on Mars.
  c) If there is water, that means there is oxygen and therefore that we can breathe on Mars.
  d) All of the above.
 $e) None of the above.

 The terrestrial (inner) planets are characterized by
  a) low density.
  b) rapid rotation.
  c) large size.
$d) rocky composition.
(*Note:* More than one answer may apply)

The Jovian planets are characterized by their
  a) high density.
  b) slow rotation.
  c) small size.
$d) composition - mostly hydrogen and helium.
(*Note:* More than one answer may apply)

The Jovian planets have retained most of their atmospheres because

a) they are further from the Sun's gravitational pull than the terrestrial planets.
b) they are very warm and massive.
$c) they are very cold and massive.
d) the planet's rings "fence in" the planetary atmosphere.

What is responsible for heating the terrestrial planets from inside?
a) temperature
b) solar radiation
$c) radioactivity
d) fusion

A planet can never be seen at opposition whose orbit relative to the Earth's has a
$a) smaller radius.
b) larger radius.
c) longer year.
d) greater ellipticity.

The synodic period of a planet is the time during which a body in the Solar System
makes one orbit of the Sun relative to the Earth, that is, returns to the same
elongation. For a superior planet to the Earth, as the synodic period increases, its
sidereal period must
$a) decrease.
b) increase.
c) stay the same.

Consider the planets closer to the Sun than Jupiter. Of these planets, the two with
the largest elongations have the _____ distances from the Sun.
$a) largest
b) smallest
c) the same
d) not enough information

If Venus is observed to have a crescent phase while its elongation angle (away
from the Sun) is 20 degrees, it must be _____ to/from the Earth than the Sun is.
$a) closer
b) farther
c) equally distant
d) Not enough information given.

Venus is unusual because of its retrograde rotation. This means that Venus
a) rotates much more slowly than the other planets.
b) does not always rotate with its axis at the same angle to the plane of the
ecliptic.
$c) rotates in an opposite direction compared to most other planets in the Solar
System.

   d) changes its direction of rotation.

The retrograde loops of Saturn are smaller than those of Jupiter because
   a) Saturn moves more slowly in its orbit.
  $b) Saturn is farther away, and loop size is largely determined by the planet's parallax.
   c) the motions of Saturn are more gradual since it has a lower temperature.
   d) the older parts of the Solar System move in smaller circular motions.

Why do Mercury and the Moon have almost no atmosphere?
  $a) The gravity at their surfaces is low, so most gas molecules travel fast enough to escape the planet.
   b) The temperature at their surfaces is high, so most gas molecules travel fast enough to escape the planet.
   c) The only gas molecules that they had originally were low in mass, so that they were immediately able to escape.
   d) They are both highly reflective.

If it its mass were closer to those of Earth and Venus, Mars might have had
   a) an active interior.
   b) a larger atmosphere than Earth.
   c) abundant water, but larger ice caps.
   d) a magnetic field.
  $e) all of the above.

Which of the following assumptions of the original solar nebula hypothesis is probably wrong, in light of the evidence provided by extrasolar planets so far?
   a) Planets as massive as the Jovian planets must be composed primarily of hydrogen and helium.
   b) Jovian planets are likely to form relatively far from the star.
   c) Centrifugal forces (due to increasing orbital speeds) prevent the gas in the solar nebula from being pulled into the star.
  $d) Planetary orbits do not shrink after the planet forms.

For a superior planet in our Solar System, the correct sequence of events as seen from Earth is
   a) opposition, eastern quadrature, western quadrature, conjunction.
   b) opposition, western quadrature, conjunction, eastern quadrature.
  $c) opposition, eastern quadrature, conjunction, western quadrature.
   d) conjunction, eastern quadrature, western quadrature, opposition.

For an inferior planet in our Solar System, the correct sequence of events as seen from Earth is
  $a) superior conjunction, greatest eastern elongation, inferior conjunction, greatest western elongation.

b) inferior conjunction, greatest eastern elongation, inferior conjunction, greatest western elongation.

c) superior conjunction, greatest western elongation, inferior conjunction, greatest eastern elongation.

d) inferior conjunction, greatest eastern elongation, opposition, greatest western elongation.

As seen from Earth, a superior planet can never be observed at
a) superior conjunction.
b) quadrature.
$c) inferior conjunction.
d) opposition.
e) none of these.

When at greatest western elongation, Venus may be observed
$a) West of the Sun in the morning sky.
b) West of the Sun in the evening sky.
c) East of the Sun in the morning sky.
d) East of the Sun in the evening sky.

When at greatest eastern elongation, Venus may be observed
a) West of the Sun in the morning sky.
b) West of the Sun in the evening sky.
c) East of the Sun in the morning sky.
$d) East of the Sun in the evening sky.

The greatest elongation of Mercury is about
a) 45 degrees.
b) 90 degrees.
$c) 28 degrees.
d) 23.5 degrees.

When Jupiter is at quadrature, the angle between Jupiter and the Sun as seen from Earth is
a) 0 degrees.
$b) 90 degrees.
c) 180 degrees.
d) 45 degrees.

When Saturn is at conjunction, the angle between Saturn and the Earth as seen from the Sun is
a) 0 degrees.
b) 90 degrees.
$c) 180 degrees.
d) 45 degrees.

When Mercury is at greatest elongation, the angle between Earth and the Sun as seen from Mercury is
   a) 0 degrees.
   $b) 90 degrees.
   c) 180 degrees.
   d) 45 degrees.

Mars is
   a) observed at opposition every year.
   b) observed at opposition twice a year.
   $c) observed at opposition every two years.
   d) never observed at opposition.

ASTEROIDS, METEORS AND COMETS

Asteroids are unlikely to be fragmented planets because
   $a)   the total mass of asteroids is small.
   $b)   no known mechanism exists to disrupt a planet.
   c)   asteroid orbits are in the ecliptic.
   d)   asteroids are typically only a few kilometers across.
   e)   asteroids can have different colors.
(*Note:* You may choose more than one answer.)

Comets could be (and were) proven to be beyond the Moon by observing
   a)   a collision of one with the Moon.
   b)   the relative size of the comet tail and the Moon.
   c)   a parallax larger than the Moon's.
   $d)   a parallax smaller than the Moon's.

Holga and Hilda are 2 asteroids. They have the same size and shape and are at the same distance from Sun. Holga is icy with an albedo of 0.9. Hilda is rocky with an albedo of 0.1. Which is brighter in visible light?
   $a) Holga
   b) Hilda
   c) They are equally bright.

Holga and Hilda are 2 asteroids. They have the same size and shape and are at the same distance from Sun. Holga is icy with an albedo of 0.9. Hilda is rocky with and albedo of 0.1. Which is brighter in infrared light?
   a) Holga
   $b) Hilda
   c) They are equally bright.

Long-period comets have orbits

a) the same as the orbits of short-period comets.
b) that are circular.
c) always in the ecliptic.
$d) randomly oriented with respect to the ecliptic.
e) of low eccentricity.

The location of the Oort cloud of comets is
a) near Pluto's orbit.
b) between the orbits of Mars and Jupiter.
$c) between 1/6 to 1/2 the distance to the nearest star.
d) close to the nearest star.
e) we have no idea since it has never been detected.

On a given day, one is likely to see the most meteors ("shooting stars")
$a) after midnight.
b) before midnight.
c) during the afternoon.
d) it doesn't depend on the time of day.

The tail of a comet is generally directed
$a) away from the Sun because of the solar wind and radiation pressure.
b) opposite the direction of motion as the comet passes through interplanetary matter.
c) toward the Sun because of the Sun's gravitational force.
d) along the comet's magnetic field lines.

Which of the following could cause a meteor shower?
a) A meteoroid hits a cloud in the atmosphere, and creates a thunderstorm.
$b) Earth crosses the debris-filled orbit of a comet.
c) Asteroids in the same orbit as the Earth.
d) A small constellation of dying stars disintegrates.

An experiment is done in the Earth's upper atmosphere in which two kinds of micrometeorites are collected: cometary and asteroidal particles. Which of the following statements is true about these particles?
a) All particles have about the same velocity.
b) Asteroidal particles typically move faster.
$c) Cometary particles typically move faster.
d) These particles are all moving slower than Earth's escape velocity.
e) Two of the above.
[*Optional hint*: If you drop a ball from a height of 5 feet, will it strike the ground with a faster or slower velocity than a ball dropped from a height of 20 feet?]

Which of Jupiter's major satellites probably suffers from the highest velocity impacts?

$a) Io. (Mass 9×10^{22} kg, Distance 421,600 km)

  b) Europa. (Mass 4.8×10^{22} kg, Distance 670,900 km)

  c) Ganymede. (Mass 1.48×10^{23} kg, Distance 1,070,000 km)

  d) Callisto. (Mass 1.08×10^{23} kg, Distance 1,883,000 km)

  e) All suffer the same impact velocities.

THE SUN

Compared to the Sun, most other stars in the Milky Way Galaxy are
  a) as small relative to the Sun as they appear in the sky.
$b) smaller.
  c) about the same size.
  d) much larger.

The Sun's luminosity comes primarily from
  a) chemical burning.
  b) the mechanical energy of turbulence.
$c) nuclear fusion.
  d) gravitational contraction.
  e) all of the above are comparable in importance.

The energy emitted by the Sun is produced
  $a) in a very small region at the very center of the Sun.
  b) uniformly throughout the whole Sun.
  c) throughout the whole Sun, but more in the center than at the surface, as $1/r^2$
  d) from radioactive elements created in the Big Bang.

The photosphere (the visible surface) of the Sun is like
  a) the surface of the Earth; you could stand on it, if you could survive the heat.
  b) the surface of the ocean; you couldn't stand on it, but you would clearly be able to detect differences above and below it.
  $c) an apparent surface; you would notice very little change as you go through it, as when you fly through a cloud.
  d) the surface of a trampoline; you could land on it, but the intense pressure would push you away again.

Sunspots appear dark because they are
  a) holes in the photosphere through which you can see deeply into the stellar interior.
  $b) a bit cooler and thus dimmer than the rest of the photosphere.
  c) large opaque structures that block light from the glowing solar surface.

d) causing retinal damage.

The 11 year solar cycle is NOT followed by the
   a) number of sunspots on the Sun.
   b) typical latitude of sunspots on the Sun.
   c) rate of solar flares.
   d) incidence of strong aurora on the Earth.
   $e) None of the above.

The chemical composition of the Sun 3 billion years ago was different from what it is now in that it had
   $a) more hydrogen.
   b) more helium.
   c) more nitrogen.
   d) molecular hydrogen.

Inside a star the mass of our Sun, energy is transported from the deep interior out toward the surface by a process most like the process that
   a) makes a blacktop hot in the summertime.
   b) heats the coil on an electric stove.
   c) makes hot water boil.
   $d) makes your hand feel heat when you put it near a candle flame.

The temperature in and around the Sun
   a) drops continuously as you move outward.
   b) rises continuously as you move outward.
   $c) drops as you move from the center to the photosphere, then rises above the photosphere.
   d) drops as you move from the center to the photosphere, then rises above the photosphere.

The light from the East limb (edge) of the Sun is blueshifted and the light from the West limb is redshifted. This is because
   a) different kinds of atoms emit light at the opposite edges.
   $b) the Sun is rotating.
   c) the distance from the Sun to the Earth changes.
   d) the two sides of the Sun are at different temperatures.

The energy of a photon emitted by thermonuclear processes in the core of the Sun takes thousands or even millions of years to emerge from the surface because
   a) it is circling in the gravitational field of the Sun.
   b) it loses energy due to convection, conduction, and radiation.
   c) of the Sun's enormous radius.
   $d) it is absorbed and re-emitted countless times along the way.

If the center of the Sun could be heated slightly, the nuclear reactions would occur faster and hence release more heat, so the Sun's core would
   a) collapse.
$b) expand and hence cool back to its previous temperature.
   c) expand and hence heat up even more.
   d) explode.

Tremendous pressure is created at the Sun's center due to its own gravity. The Sun is kept from collapsing by
   a) neutrinos and other particles generated by nuclear fusion.
   b) a hard inner core.
$c) thermal (gas) pressure generated by nuclear fusion.
   d) thermal (gas) pressure left over from the formation of the Sun.

BASIC STELLAR PROPERTIES

Stars are
   a) solid.
   b) liquid.
$c) gaseous.
   d) mostly carbon, oxygen, nitrogen, and iron.
   e) both c and d.

If the Sun were the size of a tennis ball (6.35 cm in diameter) then the red giant star Betelgeuse would have a diameter of
   a) 2.3 cm, about the size of a Ping-Pong ball.
   b) 23 cm, about the size of a basketball.
$c) 23 m, larger than the equivalent size of Mars' orbit.
   d) 23 km.

Most of the brightest stars in the sky are
   a) relatively hot main-sequence stars that are relatively close to the Sun.
   b) relatively cool giant stars that are relatively close to the Sun.
   c) relatively cool main-sequence stars that are relatively far from the Sun.
   d) relatively cool main-sequence stars that are relatively close to the Sun.
$e) giant stars and relatively hot main sequence stars.

Two stars have the same chemical composition, spectral type, and luminosity class, but one is 10 light years from the Earth and the other is 1000 light years from the Earth. The farther star appears to be
   a) 10 times fainter.
   b) 100 times fainter.
$c) 10,000 times fainter.
   d) 100,000,000 times fainter.

e) the same brightness since the stars are identical.

Two stars have the same surface temperature, but the radius of one is 100 times that of the other. The larger star is
   a) 10 times more luminous.
   b) 100 times more luminous.
   $c) 10,000 times more luminous.
   d) 100,000,000 times more luminous.
   e) the same luminosity.

Which property of a star would not change if we could observe it from twice as far away?
   a) angular size.
   $b) color
   c) flux
   d) parallax
   e) proper motion

Suppose that two stars are at equal distance and have the same radius, but one has a temperature that is about twice as great as the other. The flux from the hotter star is
   $a) about 16 times greater.
   b) about 16 times less.
   c) about 20 per cent greater.
   d) about 20 per cent less.
   e) Not enough information given to answer the question.

Star A has a radius R and temperature T. At the same distance from Earth, star B has a radius 4R and temperature T/2. Which star appears to be brighter?
   a) A
   b) B
   $c) Both stars appear to have the same brightness.
   d) Cannot tell from information given.

Two stars have the same temperature, but the radius of one is twice that of the other. How much brighter is the larger star?
   a) the same because luminosity depends only on temperature.
   b) 2 times
   $c) 4 times
   d) 8 times
   e) 16 times

Consider two stars in constellation Pasta:
      Alpha Tortellini – bright in UV, dim in IR
      Beta Ziti – dim in UV, bright in IR

There is little or no dust along the line of sight to Pasta. Which star is hotter?
$a) Alpha Tortellini
  b) Beta Ziti
  c) Both are the same temperature.

A main sequence star is stable because of self-regulation;
  a) the nuclear reaction rate does not depend on temperature.
  b) a slight contraction decreases the nuclear reaction rate.
$c) a slight contraction leads to higher gas pressure.
  d) a slight contraction leads to lower internal pressure.

STAR FORMATION

The material that makes up the Sun was once part of
  a) the Big Bang.
  b) another star.
  c) a molecular cloud.
  d) a protostar.
$e) all of the above.

If the material in the primordial Solar System retained its angular momentum as it collapsed to form the Sun, the Sun's rotation rate should be
  $a) fast (less than a week).
  b) moderate (a week to a month).
  c) slow (more than a month).
  d) zero (non-rotating).

If an interstellar cloud contracts to become a star, it is due to which force?
  a) electromagnetic
  b) nuclear
  $c) gravitational
  d) centrifugal

Stars are formed from cold interstellar gas clouds made up of
  a) atomic gas of mostly hydrogen, oxygen, carbon and nitrogen.
  $b) molecular hydrogen gas and dust grains.
  c) some hydrogen gas, comets and asteroids.
  d) stars are formed from very HOT gas.

Evidence for an association between interstellar dust and star formation comes from
  $a) dark clouds that attenuate the light from background stars.
   b) emission at radio wavelengths of 21 cm.
  $c) stars that appear redder than expected from their spectral type.

d) bright nebulae with redshifted emission lines.

As a star first begins to condense from dust and gas clouds, it emits primarily in which wavelength regime?
 a) X-ray
 b) ultraviolet
 c) visual
$d) infrared
 e) radio

The initial collapse of the solar nebula may have been initiated by a nearby supernova explosion because
 a) there is no other known mechanism.
$b) pre-stellar clouds are normally stable.
$c) certain odd isotopes are found in the Solar System.
 d) only 90% of atoms in the Solar System are hydrogen.
(*Note:* More than one answer may apply.)

The factors that tend to resist collapse in a gas cloud are
   I. the magnetic field of the cloud
   II. the presence of supernovae near the gas cloud
   III. the existence of molecules in the gas
   IV. heat in the gas
   V. the rotation of the cloud
 a) I, III, V.
$b) I, IV, V.
 c) II, III, IV.
 d) II, IV, V.
 e) III, IV, V.

We know that low mass proto-stars have strong stellar winds because
 a) they are collapsing in a stable fashion.
 b) they have thick circumstellar disks.
$c) the Sun was once a proto-star and has a wind now.
$d) proto-stars have gas jets that drive massive outflows.
(*Note:* More than one answer may apply.)

The nebulae around T Tauri stars are shaped into disks because of the same process that causes dough to become flat when it is
  $a) spun in the air like pizza dough.
   b) rolled with a pin like pizza dough.
   c) squashed between plates like dough for a burrito.
   d) baked on a flat sheet like a cookie.

What important event occurred while the Sun was in its T Tauri phase (a phase young stars go through in which they have a strong stellar wind similar to the solar wind)?

    $a) Gas and dust remaining in the Solar System were blown away.
    b) Comets were pushed out of the Solar System.
    c) The planets formed.
    d) The Sun shrank to its present size.

ENERGY GENERATION IN STARS
AND STELLAR EVOLUTION

Energy Generation in Stars

Why does fusion of hydrogen release energy?

    a) Fusion breaks the electromagnetic bonds between hydrogen atoms, releasing energetic photons.
    $b) The mass of a helium nucleus is smaller than the mass of four protons.
    c) The mass of a helium nucleus is larger than the mass of four protons.
    d) The velocity of four protons is larger than the velocity of a helium nucleus.
    e) None of the above is true.

Why would two protons combine to form an atom of deuterium (heavy hydrogen) in the core of a star like the Sun?

    a) The electromagnetic force strongly attracts the protons.
    b) The gravitational force strongly attracts the protons.
    $c) The velocity of protons in the core of the Sun is very large.
    d) Protons never combine to form deuterium in the core of the Sun.
    e) Both a and c.

Fusion in the core of a stable massive star cannot proceed beyond iron because

    a) it would require temperatures that even stars cannot generate.
    b) the fusion of iron nuclei is impossible under any circumstances.
    $c) iron nuclei are the most tightly bound of all nuclei so iron fusion does not release energy.
    d) the Chandrasekhar limit has been reached, so a black hole must result.

After the Sun's core hydrogen is depleted by nuclear fusion the core will consist primarily of

    a) carbon.
    b) deuterium.
    $c) helium.
    d) oxygen.

The solar corona has temperatures roughly the same as temperatures in the Sun's core, where nuclear fusion takes place. Fusion doesn't occur in the corona because
$a) the density in the corona is too low.
   b) the corona has too many free electrons.
   c) atoms in the corona are mostly ionized.
   d) the corona has more heavy atoms than the core.
   e) Two of the above.

Fusion in the core of a main sequence star changes the chemical composition in the core. What happens to the chemical composition outside the core?
   a) We have no way of finding out.
$b) The chemical composition outside the core changes very little.
   c) The same changes occur outside the core as within the core.
   d) Hydrogen becomes more abundant outside the core.

Which of the following would lengthen the amount of time a star is able to fuse hydrogen at its center?
   a) The temperature in the core of the star is increased.
$b) The gas in the core of the star is enriched in hydrogen.
   c) The mass of the star is increased.
   d) none of the above

About 25% of the mass of a newborn star is initially in helium. Why doesn't helium also fuse on the main sequence?
      I. Helium nuclei travel more slowly on average than hydrogen nuclei at any temperature
      II. Helium nuclei repel each other with more force than do hydrogen nuclei.
      III. Helium fusion requires three helium nuclei to hit each other almost simultaneously.
   a) I
   b) II
   c) III
   d) I, II
   e) II, III
$f) I, II, III

If the rate of hydrogen fusion within the Sun were to increase, the core of the Sun would
   a) contract and decrease in temperature.
   b) expand and increase in temperature.
$c) expand and decrease in temperature.
   d) stay the same size but increase in temperature.

If the temperature in the core of the Sun increased,
   a) the rate of nuclear reactions in the core would increase.

b) the radiation pressure in the core would increase.
c) the core of the star would expand.
d) the temperature in the core would decrease.
$e) all of the above would occur.

If the core of the Sun were somehow kept extremely cold for a long time, the Sun would
   a) freeze into ice.
   b) expand to about the radius of Earth's orbit.
$c) collapse to about the size of the Earth.
   d) collapse to about the size of a large city.

A celestial body growing by accretion of material must surpass a certain mass before hydrogen fusion begins in the core, making it a star. If the strength of the charge on the proton were to be increased, then that minimum mass
$a) must increase.
   b) must decrease.
   c) would not change.

Stellar Evolution

If two stars are on the main sequence, and one is more luminous than the other, we can be sure that the
   a) more luminous star will have the longer lifetime.
   b) fainter star is the more massive.
$c) more luminous star is the more massive.
   d) more luminous star will have the redder color.

When a star becomes a red giant, it becomes much brighter because it is
   a) moving closer to us.
   b) losing its outer envelope.
   c) fusing iron in its core.
$d) increasing in size.

As a one solar mass star evolves into a red giant, its
   a) surface temperature and luminosity increase.
   b) surface temperature and luminosity decrease.
   c) luminosity decreases while the surface temperature increases.
$d) luminosity increases while the surface temperature decreases.

After hydrogen fusion stops in the core of a star, the core
   a) cools and the star as a whole expands.
   b) cools and the star as a whole contracts.
$c) heats and the star as a whole expands.

d) heats and the star as a whole contracts.

A star is burning hydrogen to helium in its core and has 10 times the mass of the Sun. Which of the following are true?
   a) The surface temperature of the star is smaller than that of the Sun.
   b) The star is redder, more luminous, and larger than the Sun.
   $c) The star is bluer, larger, and more luminous than the Sun.
   d) both a and b.
   e) both a and c.

A star evolves off the main sequence when
   a) nuclear reactions begin in the core of the star.
   $b) hydrogen is exhausted in the core of the star.
   c) hydrogen is exhausted everywhere in the star.
   d) helium is exhausted in the core of the star.

Giant stars are more rare than main sequence stars because
   a) they do not form as often as main sequence stars.
   b) giant stars are unstable.
   $c) the giant stage is very short compared to the main sequence stage.
   d) elements heavier than helium are relatively rare.

Which of the following stars is probably oldest?
   a) a 1 solar mass main sequence star
   $b) a 1 solar mass white dwarf
   c) a 10 solar mass main sequence star
   d) a 10 solar mass red giant

After the first 20 billion years of the Universe, the Sun will have evolved through red giant, main sequence, and white dwarf stages. Assuming a current radius of 1, what radii will the Sun have during these phases, in correct order?
   $a) 1, 400, 0.01
   b) 0.01, 1, 400
   c) 1, 1, 1
   d) 1, 0.01, 400

After a star has evolved into a red giant, hydrogen burning
   a) ceases completely.
   b) happens only in the center of the star.
   $c) happens only in shells outside the core of the star.
   d) happens only during novae.

Which of these stars will end its main sequence lifetime most rapidly?
   $a) A very massive star, since more massive stars consume their hydrogen more rapidly.

b) A low-mass star, since less massive stars have less hydrogen to burn.

c) A star like the Sun, since the combination of fuel-use rate and available fuel amount peaks near 1 solar mass.

d) none of the above, since all stars have main sequence lifetimes of about 10 billion years.

The event that marks the end of a star's evolutionary life before becoming a white dwarf is
   a) a nova.
   $b) a planetary nebula.
   c) the exhaustion of hydrogen in the core.
   d) a helium flash.

The force of gravity acts to
   a) make a star larger.
   $b) make a star smaller.
   c) make a star cooler.
   d) none of these.

When during a star's evolution its core gets smaller, the rest of the star typically
   a) also gets smaller.
   b) stays the same size.
   $c) gets larger.
   d) explodes in a supernova.

Suppose in a given region of the sky, you see a red star and a blue star. The two are not parts of binary systems, and both stars look pretty typical for their colors. Which of the following is true?
   a) The red star is older than the blue star.
   b) The red star is younger than the blue star.
   $c) The stellar ages cannot be determined from the information given.

The helium burning phase for a star of a given mass is much shorter than the hydrogen burning phase primarily because
   a) the helium mass fraction in the core is less than the hydrogen mass fraction was when the star was young.
   $b) helium releases less energy per reaction than hydrogen.
   c) the star becomes a white dwarf before it can use most of its helium.
   d) the temperature never rises high enough for complete helium burning.

A star remains at constant size and temperature for a long period of time. Which of the following is most likely to be true? The star generates
   a) more energy than it radiates into space.
   $b) about as much energy as it radiates.
   c) less energy than it radiates into space.

You are an immortal alien being, hiding in the photo archive room of a library on Earth. You can best learn about the life cycles of people by bringing home the drawer filled with photographs of
   a) individuals.
   $b) crowds on the street.
   c) people lined up at the voting both.
   d) doctors.

Black Holes

A black hole is best defined as
   a) a star that sucks all matter into itself.
   b) a window to another Universe.
   $c) any object that is smaller than its event horizon.
   d) the final result of all stellar evolution.

Which of the following can escape from inside the event horizon of a black hole?
   a) particles of matter
   b) particles of antimatter
   c) visible light
   d) X-rays
   $e) None of the above.

Isolated black holes slowly evaporate because they slowly leak mass via
   a) faster-than-light particles that can escape.
   $b) virtual particles that form near the event horizon.
   d) nuclear fusion near the event horizon.
   c) holes in the event horizon.

The best current theories about an isolated black hole suggests that radiation is
   a) impossible because it is a hole.
   b) impossible because the escape velocity exceeds the speed of light.
   c) lower from smaller black holes.
   $d) higher from smaller black holes.

BINARY STAR SYSTEMS

General Concepts of Binary Stars

Two stars are in a binary star system. Star A is 5 solar masses. Star B is 2 solar masses. They are separated by a distance of 200 AU. The period of star A around the center of mass is _____ the period for star B around the center of mass.

a) larger than
b) smaller than
$c) the same as

Considering a representative sample of 100 star systems in our Galaxy,
  $a) stars in binary systems would be more common than single stars.
  b) about half of the stars would be in resolved visual binaries.
  c) about half of the stars would be eclipsing binaries.
  d) most would be single stars.

In general, the orbits of binary stars are
  a) circular.
  b) of high eccentricity.
  $c) oriented more or less randomly in space.
  d) tilted perpendicular to the line of sight.

Binary Classifications and Uses

The most important feature of binary stars is that they enable us to determine stellar
  a) temperatures.
  b) pressures.
  c) compositions.
  $d) masses.

An astrometric binary is known to have an unseen companion whose mass is about equal to the Sun's. The companion is most likely a
  a) black hole.
  b) pulsar.
  c) supergiant.
  $d) white dwarf.

If we know the distance to an eclipsing binary system, we can determine the member stars'
  a) temperatures.
  b) radii.
  c) masses.
  d) luminosities.
  $e) All of the above.

In a certain spectroscopic binary, one star's spectral lines vary in wavelength twice as much as the other. The star whose lines move more is
  $a) half the mass of the other.
  b) twice the mass of the other.

c) twice as far from Earth as the other.
d) twice as luminous as the other.

If we observe the orbit of a binary star system face-on, it is most likely to be detected as a binary
$a) visually (separation in the plane of the sky).
b) spectroscopically (periodic velocity variation).
c) astrometrically (cyclic motion in the plane of the sky).
d) by eclipses (periodic photometric variability).

If we observe more distant binary star systems, we are most likely to discover their binarity
$a) spectroscopically (periodic velocity variation).
b) visually (spatially resolved).
c) astrometrically (cyclic motion in the plane of the sky).
d) by eclipses (periodic photometric variability).

Compared to other types of binaries, a binary star system discovered as a visual binary is likely to have relatively
a) high orbital velocities.
$b) large separation, long period.
c) large separation, short period.
d) small separation, long period.
e) small separation, short period.

For a long-period spatially resolved (visual) binary, the observations that imply it is a binary system are the
a) velocity curves of the components.
b) differing brightnesses (eclipses).
c) Doppler velocities of the components.
$d) small separation and common proper motion.

When compared to visual, spectroscopic, or eclipsing binaries, optical doubles are not true binaries because
a) they are only binaries in the optical.
$b) they are not gravitationally bound.
c) their radial velocities do not vary.
d) they have no proper motions.

If we see the orbital plane of two distant tightly bound stars nearly edge-on, we are most likely to discover the binary system as a
$a) spectroscopic and eclipsing binary.
b) spectroscopic and long-period binary.
c) visual and astrometric binary.
d) visual and eclipsing binary.

Which of the following is most likely?
  a) A spectroscopic binary has a period of over a decade.
  b) A spectroscopic binary is also a visual binary.
  c) A system that produces a nova is also a visual binary.
  $d) An eclipsing binary is also a spectroscopic binary.
  e) An eclipsing binary is also a visual binary.

Binary System Light Curves

Flat-bottomed minima in the light curve of an eclipsing binary imply
  a) a large temperature difference between the components.
  b) eccentric orbits.
  c) partial eclipses.
  d) tidal distortion.
  $e) total and annular eclipses.

If a spectroscopic binary is also eclipsing, the maximum Doppler shift in the spectral lines occurs
  a) each time an eclipse takes place.
  b) every other time an eclipse takes place.
  $c) halfway between eclipses.
  d) between eclipses, but the timing depends on inclination.

In an eclipsing binary, the deeper (more dimmed) eclipse occurs when the _____ star is being eclipsed.
  $a) hotter
  b) larger
  c) more luminous
  d) more massive
  e) smaller

For an eclipsing binary, from just the light curve one can find the
  a) masses of both stars.
  b) luminosities of both stars.
  c) relative masses and diameters of the stars.
  d) relative masses of the stars.
  $e) relative sizes of the stars.

STELLAR POPULATIONS

Constellations

Constellations in astronomy are

a) physical groupings of genuinely associated stars.
b) swarms of planets or asteroids.
c) the most accurate way to predict the future.
$d) arbitrary but useful subdivisions of the sky.

Which of the following statements are true?
a) Stars aren't moving with respect to each other, although if they are nearby then they can appear to move due to parallax.
b) Stars like the Sun emit only yellow light, while cooler stars emit only red light.
c) Most of the 20 closest stars to the Sun are among the 20 brightest stars in the sky.
d) All of the above are true.
$e) None of the above is true.

Suppose you measure the parallax of each star in the constellation Taurus (or any other constellation you might choose). Which of the following is the most likely?
a) The stars all have the same parallax since we see them together in the same constellation.
b) The stars all have nearly the same parallax since they are moving together through space.
c) None of them has a measurable parallax since they are mostly within our own Solar System.
$d) They may have significantly different parallaxes.
e) We cannot measure their parallaxes since they are all moving toward our Sun.

Star Clusters

If you find a cluster of very old stars, you would expect the cluster to appear
$a) very concentrated toward its center.
b) very spread out.
c) shaped like a disk.
d) bipolar in shape, like a barbell.

As a star cluster evolves over time,
a) the composition of the stars in the cluster changes from metal-poor to metal-rich.
$b) more and more main-sequence stars become red giants.
c) the cluster changes its location in the galaxy, getting closer to the spiral arms.
d) the shape of the cluster changes to a disk shape.

In a color-magnitude diagram of a star cluster, the blue end of the main sequence is useful for defining the age of the cluster because
a) blue stars are not affected by extinction and reddening by dust.

$b) stars just slightly brighter and redder are just now evolving off the main sequence to become giants.

   c) older, metal-poor stars are blue.

   d) the hottest stars are the oldest stars in a cluster.

The difference in apparent magnitude of the main sequences of stars from two clusters in a color-magnitude diagram tells us

   a) how much each cluster has evolved.

   b) if one cluster is "open" and the other is globular.

   c) the relative ages of the clusters.

$d) the relative distances of the clusters.

Star clusters are useful to stellar astronomers because the clusters contain stars that

   a) are all about the same age.

   b) span a wide range of ages.

   c) are all at the same stage of stellar evolution.

$d) are all about the same age and distance.

Because stars in clusters all have similar age and distance, the main underlying physical cause of their different appearances is their

   a) color.

   b) radius.

$c) mass.

   d) chemical composition.

   e) temperature.

When making a color-magnitude diagram of a star cluster, we can use apparent brightness instead of intrinsic luminosity because

$a) all stars in the cluster are at about the same distance.

   b) all stars in the cluster have about the same age.

   c) most stars in the cluster are on the main sequence.

   d) reddening by interstellar dust does not affect clusters.

   e) the cluster stars have similar chemical compositions.

Young stars contain heavy elements ("metals") while old stars do not because

   a) young stars are hotter than old stars.

   b) young stars are more massive than old stars.

   c) old stars have used up all their metals in producing energy.

$d) the gas out of which young stars were formed already contained metals produced by older stars.

OUR GALAXY

The position of the Sun in the Milky Way Galaxy is best described as

$a) in the disk, slightly more than halfway out from the center.
 b) very close to the center.
 c) in an open cluster in the disk.
 d) in a globular cluster in the halo.

Trigonometric parallax, the apparent motion of stars due to the Earth's annual motion around the Sun, can be used to study the structure of our galaxy
   a) within the Solar System.
   b) up to about 1 parsec (3.26 light years) from the Sun.
 $c) within about 100 parsecs of the Sun.
   d) within about 100 parsecs of the galactic center.

The disk of stars that forms the major component of the Milky Way Galaxy has its shape due to
   a) gas pressure from outside the galaxy.
   b) collapse due to self-gravity.
   c) rotation alone.
 $d) rotation combined with self-gravity.
   e) the cosmological expansion of the Universe.

The thickness of our galaxy's disk is determined by the
   a) circular speed of stars around the galaxy.
   b) random motion of stars in the plane of the disk.
 $c) random motion of stars perpendicular to the disk.
   d) the amount of matter in nucleus of the galaxy.

When we observe stars near the center of the Milky Way Galaxy, we detect light that was emitted from those stars about
   a) 200 million years ago.
 $b) 25,000 years ago.
   c) 8 years ago.
   d) 8 minutes ago.
   e) when the galaxy was formed.

An observer far outside our galaxy would best describe our galaxy and the Sun's position in it as a
 $a) disk of stars with our Solar System near the edge.
   b) disk of stars centered on our Solar System.
   c) disk of stars with a bulge containing our Solar System.
   d) sphere of stars centered on our Solar System.
   e) sphere of stars with our Solar System near the edge.

The mass of our galaxy is best found by
   a) counting the number of stars in the sky.
   b) counting the star clusters in the sky.

c) counting the hot, massive main sequence stars.
d) radio measurements of the amount of interstellar hydrogen.
$e) measuring the rotation of the Galaxy.

You could best map out the overall spiral structure of our Galaxy by finding
$a) young open clusters and neutral atomic hydrogen gas.
b) evolved stars like planetary nebulae, RR Lyrae stars.
c) smooth, round globular clusters.
d) high velocity stars.

Choose the best evidence that the disk of the Milky Way Galaxy does NOT rotate like a solid wheel.
$a) Disk stars have Doppler shifts.
b) The brightest disk stars form spiral arm shapes.
c) Disk stars rotate twice as quickly that are twice as far from the Galactic center.
d) The rotation of disk stars around the Sun decreases with distance according to Kepler's laws.

Compared to stars like the Sun in the disk of the Milky Way, stars that populate the extended spheroidal halo of the galaxy were born
a) earlier, so have had time to accumulate more heavy elements.
b) later, so have used up their heavy elements.
$c) earlier, from more nearly primordial material, so have fewer heavy elements.
d) later, so have accumulated more heavy elements from previous generations of stars.

You observe two stars at the same distance. One is in the disk of the Milky Way, the other in a direction perpendicularly out of the disk. Chances are that the disk star will be
a) less luminous, have a smaller Doppler shift and be reddened by dust.
b) more luminous, have a smaller Doppler shift and be reddened by dust.
$c) more luminous, have a larger Doppler shift and be reddened by dust.
d) more luminous, have a smaller Doppler shift and be less reddened by dust.

A cold thin cloud of interstellar gas embedded in a hotter, thinner surrounding medium would yield 21-cm radio spectral emission lines of hydrogen showing
a) a single broadened emission line.
b) broad emission with a narrow spike-like absorption core.
c) broad absorption with a narrow spike-like emission core.
$d) a narrow spike-like emission core and low broad shoulders of emission.

The primary reason that massive O-type stars are not found in the galactic halo is because they are
a) too massive to be kicked into the halo from the disk.
b) so massive that they settle into the thinner disk.

$c) too short-lived to have persisted from halo formation until today.
 d) closer to us in the disk than in the extended halo.

One model for the formation of the Milky Way can be divided into 2 phases: a spherical gas cloud collapsed to form the stars in the Milky Way's spheroid, then rapidly rotating gas collapsed into a disk-shaped configuration of stars. Since disk stars have higher metallicity, which is most likely? Gas ejected from the
 $a) spheroid stars enriched the material now in the disk stars.
 b) spheroid stars decreased their metallicity.
 c) spheroid decreased its angular momentum.
 d) disk stars puffed out the spheroid stars into a rounder shape.

NORMAL GALAXIES

Most of the hydrogen molecules (two hydrogen atoms bound together) in the Universe are found in
 a) cool stars.
 $b) cold dense interstellar clouds.
 c) H II regions of ionized hydrogen.
 e) space between the galaxies.

If one region of the sky shows nearby stars but no distant stars or galaxies, our view is probably blocked by
 a) nothing, but directed toward a particularly empty region of space.
 b) an emission nebula of ionized gas.
 $c) an interstellar gas and dust cloud.
 d) a concentration of dark matter.

The dominant mass component in the Milky Way Galaxy interior to the orbit of the Sun is
 a) gas and dust contained in giant molecular clouds.
 b) cosmic ray particles.
 c) stars.
 $d) dark matter.

Compared to the present day Milky Way Galaxy, the Milky Way of 3 billion years ago would have had
 $a) more gas in the disk.
 b) more stars in the halo.
 c) more metal-rich stars.
 d) no Solar System.

The most effective technique to find distances to other galaxies is to look at objects in the galaxies for which we know the

$a) luminosity.
  b) color.
  c) mass.
  d) orbital velocity about the center of the galaxy.
  e) spectral type.

You are asked to determine an accurate distance to the Andromeda Galaxy. Which of the following is the best technique?
  a) Doppler shift of spectral lines
$b) period-luminosity law for Cepheid variable giants in Andromeda
  c) the Hubble Law of recession of galaxies
  d) trigonometric parallax using Earth's orbit

Suppose that the period-luminosity law for giant Cepheid variable stars in the Milky Way leaves an uncertainty of 30% in distance estimates made from the measured variability periods of Cepheid stars. Much more distant Cepheids measured in other galaxies will yield distance uncertainties that
  a) increase with distance.
  b) are at most 30%.
$c) are at least 30%.
  d) depend on the galaxy.

The Tully-Fisher method of measuring distances to spiral galaxies is based on a relation between the speed of the galaxy's rotation and its absolute luminosity. The existence of the relation suggests that
  a) distant galaxies are more luminous, with greater velocities.
  b) rotations of galaxies generate star formation and therefore light.
  c) luminosity generates rotation.
$d) luminous galaxies have more stars, and therefore more mass.

Which of the following is impossible to use in estimating distances to other galaxies. Why?
  a) globular clusters; too large and fuzzy
$b) white dwarfs; too dim
  c) pulsating variable stars; too variable
  d) supernovae; don't last long enough

Which of the following is least easily explainable as a result of interaction between galaxies?
  a) Some galaxies have long "tails" of stars.
  b) Rich, regular clusters are dominated by central giant ellipticals.
$c) Both spiral and elliptical galaxies are seen at very high redshift.
  d) Some galaxies seems to be undergoing bursts of star formation.

The nuclei of most spiral galaxies appear redder than their spiral arms because of

$a) young blue stars in the arms, and old red ones in the nuclei.
 b) emission nebulae and dust in the nuclei.
 c) receding nuclei and advancing spiral arms (Doppler shifts).
 d) nuclear reactions.

ACTIVE GALAXIES AND QUASARS

The most likely reason that clusters of galaxies have more elliptical than spiral galaxies is that in the high density cluster environment
 $a) spirals merge to form ellipticals.
 b) intracluster gas strips galaxies of the gas needed for star formation.
 c) near-misses between galaxies makes them rounder.
 d) galaxies are older and their brighter disk stars have burned out.

At high redshift, a larger fraction of galaxies are "active" (show signs of powerful luminous nuclei) than at low redshift. Therefore, we can safely say that
 a) all galaxies go through an active phase, and more galaxies in the past were active than now.
 $b) some galaxies go through an active phase and more galaxies in the past were active than now.
 c) all galaxies are either active or normal.
 d) galaxies may become active more than once in their lifetimes.

At high redshift, a larger fraction of galaxies are "active" (show signs of powerful luminous nuclei) than at low redshift. If galaxies only become active when they collide or interact with nearby galaxies, then it might be true that
 a) there were more interactions in the past, and activity fades away.
 b) the number of distinct galaxies in the Universe decreases with time.
 c) galaxies were closer together in the past.
 $d) all of the above

The large Doppler velocity widths of broad emission lines in active galaxies (Seyferts and quasars) could NOT be created by hot emitting clouds that are
 a) swirling at high velocities around a black hole.
 b) falling into the neighborhood of a black hole.
 c) being ejected into a broad cone or disk-shaped wind.
 $d) being ejected along a narrow-angled jet.

Narrow absorption lines in the spectra of distant quasars could be caused by clouds of gas on the
 a) near side of the quasar with large random velocities but small bulk velocities.
 $b) near side of the quasar with small random velocities but large bulk velocities.
 c) far side of the quasar with large random velocities but small bulk velocities.

    d) far side of the quasar with small random velocities but large bulk velocities.

Seyferts and quasars are both types of active galaxies, harboring powerful luminous nuclei. Quasar nuclei appear to be more luminous, and therefore their black holes are
    a) accreting matter at a higher rate.
    b) more massive.
    c) less obscured along our sightline.
$d) Any of the above.

The word "quasar" comes from "quasi-stellar." What makes quasars quasi-stellar is that they can have a
    a) proper motion seen between images taken at 2 epochs.
    b) Doppler velocity shift evident in their spectra.
$c) point-like appearance in an image.
    d) binary companion.

The wide variety of spectra observed from different active nuclei of galaxies appear may result from
    a) how many neutron stars they contain.
    b) the amount of dust in the Milky Way Galaxy blocking the view.
$c) the angle at which we view each nucleus.
    d) whether or not the galaxy has a close companion.

One method you could use to search for a high-mass black hole at the center of a galaxy is to look for
    a) a black dot at the galaxy's nucleus.
    b) a very high luminosity star.
$c) a very large range of Doppler shifts around the nucleus.
    d) distortion in the shapes of stars near the nucleus.

Rapid variability in the luminous nuclei of quasars is evidence that the emission region must be
$a) small.
    b) large.
    c) moving rapidly.
    d) exploding.

The powerful nuclei of quasars and Seyfert galaxies cannot be dominated by starlight because nuclear fusion in a group of stars could not account for the quasar's observed
    a) rapid variability.
    b) luminosity.
    c) compact size.
$d) All of the above.

The distance to the point-like quasars is found from
 a) comparing their apparent and absolute magnitude.
 b) the apparent magnitudes of their supernovae.
 c) their parallax measured with radio telescopes.
 $d) their redshift and the Hubble Law.

If the large redshifts of quasars were NOT caused by the cosmological expansion,
then bright quasars could be explained as
 a) distant objects that are very luminous.
 $b) nearby luminous objects exploding outward from the Milky Way.
 c) bright nearby objects severely reddened by intervening dust.
 d) distant objects severely reddened by intervening dust.

Quasars are more likely powered by accretion onto a supermassive black hole than
by stars because accretion is
 a) the only natural way to produce radio and X-ray emission.
 $b) a much more efficient means than fusion of extracting energy from matter.
 c) possible in the early Universe before stars even formed.
 d) responsible for destroying and engulfing stars.

COSMOLOGY

The Hubble Expansion

Your job is to compile a representative catalog of galaxies. Assuming our region of
the Universe is typical, the best criterion to use to decide whether to include
galaxies in the catalog is to include all galaxies on the sky with
 a) magnitudes brighter than some chosen limit.
 b) apparent diameters larger than some chosen limit.
 $c) recession velocities less than some chosen limit.
 d) surface densities of stars larger than some chosen limit.

Suppose the Universe were not expanding, but was in some kind of steady state.
How should galaxy recession velocities correlate with distance? They should
 a) be directly proportional to distance.
 b) reverse the trend we see today and correlate inversely with distance.
 c) show a scatter plot with most recession velocities positive.
 $d) show a scatter plot with equal numbers of positive and negative recession
velocities.

The blueshift exhibited by some nearby external galaxies lends support to the
 a) Big Bang model, because those galaxies are expanding towards us.
 $b) Local Group concept, because nearby galaxies interact gravitationally.

c) evolution theory, because galaxies change color as they evolve.

d) oscillating Universe theory, because some galaxies are moving together already.

Suppose you've accepted that the Universe is expanding, and will always expand. You must then accept that

a) there is either no center to the Universe, or we are at the center.

b) the Universe is a one-shot deal.

c) either the average distance between galaxies always grows, or the distance between stars in the galaxies grows.

$d) either the density of the Universe always decreases or new matter must be continuously created.

The observed redshifts of galaxies mean that

a) we are at the center of an expanding Universe.

b) gravity can never overcome the expansion left over from the Big Bang.

$c) photons lose energy traveling through space.

d) galaxies are growing smaller.

If the speed of expansion of the Universe is increasing, then redshift-based estimates of the look-back time to distant galaxies based on a steady expansion rate have been

a) too small.

$b) too large.

c) just fine.

If the Universe is expanding, won't the Solar System eventually expand apart?

a) The Solar System may actually be shrinking now, which makes the Universe LOOK like it's expanding.

$b) No, its gravity holds it together.

c) No, because there is no planetary redshift.

d) Eventually, but only after a very long time.

Suppose the Hubble Constant were measured and found to be twice as large as it is now believed to be. The implied maximum age of the Universe in a Big Bang model would be

$a) halved.

b) the same.

c) doubled.

d) squared.

The Hubble Age of the Universe, $1/H$, represents how long ago the Big Bang happened, based on its current rate of expansion. The Hubble Age must

a) ignore the gravitational interaction of matter.

$b) be a maximum possible age.

c) be a minimum possible age.

If the Hubble Constant, *H*, is larger at great distances, then the
   a) Universe must be older than we suspect.
   b) matter in the Universe is not important to its motion.
   $c) expansion is slowing.
   d) all of the above.
   e) none of these.

There must be some large distance (the Hubble Length $D = c/H$) that is too far away for light to have reached us during the age of the Universe. The expansion velocity relative to us at that distance must be
   a) zero.
   b) infinite.
   c) less than the speed of light.
   $d) the speed of light or greater.

Cosmology

Which of the following observations about the nature of the Universe can be made with only a small telescope?
   a) The Universe is expanding.
   b) Most of the matter in the Universe does not emit light.
   $c) Luminous matter in the Universe occurs in clumps rather than being evenly distributed.
   d) There is background radiation in all directions that came from the Big Bang.

We only observe events that happened in the past because
   a) it takes time to analyze the data.
   $b) the speed of light is finite.
   c) the Universe is very old.
   d) our telescopes are not yet large enough.

The look-back time to an object is the number of years between when the object emitted the light we see and the present. What piece of information surely does NOT affect the look-back time for a distant object?
   a) its distance.
   b) the speed of light.
   c) the rate of expansion of the Universe.
   $d) the wavelength of light being observed.

The cosmological principle enables astronomers to generalize from what they observe in the nearby Universe to the properties of the Universe as a whole. The principle means that no matter where you are in space, you should see that

a) galaxies are all moving away from the same point.
b) the Universe does not change with time.
$c) space looks approximately the same in all directions.
d) every region of space is unique.

The night sky is relatively dark because the Universe
a) is filled with dust.
b) is mostly empty.
c) is very large.
$d) has a finite age.

Olber's Paradox asks why the night sky is dark, when every line of sight must eventually fall on a star. Which of the following reasons would best explain the darkness at night? It is because the Universe is
a) infinite and mostly empty.
b) clumpy, so not every sightline intercepts a star.
c) expanding, so distant stars are redshifted.
$d) young, so there are only stars to a finite distance.

Suppose the Universe is closed and we observe galaxies after the Universe has already begun to contract. What would the Hubble Diagram plot of velocity versus distance look like? Galaxies would appear to have velocities that are
a) positive (receding) and still proportional to their distance.
b) inversely proportional to their distance.
$c) negative nearby, but positive at large distances.
d) positive nearby but negative at large distances.

The Steady State theory of cosmology holds that the Universe is expanding, but new matter is created so that the appearance of the Universe does not change. This theory would imply that
a) there are other galaxies in the Universe that are several times as old as our own Milky Way.
b) there was no Big Bang.
c) the space density of quasars should look the same at all redshifts.
$d) all of the above.

Suppose that the Universe is static and every object in it was created simultaneously, 13 billion years ago. Which of the following statements is true of an object that is 14 billion light years away?
a) It shows that the Universe has to be infinite.
b) Its light first reached the Earth 1 billion years ago.
c) We will never be able to see it.
$d) We will not see it until 1 billion years from now.

If the Big Bang cosmology is correct, then

a) the most distant galaxies we observe are seen now as they will be in the future.
  $b) nearby galaxies are seen at a more advanced stage of evolution than distant galaxies.
  c) matter is being continually created.
  d) the universal expansion must be slowing down.

If the Big Bang cosmology is correct *and* there is not enough mass to close the Universe, then
  a) more Big Bangs will occur in the distant future.
  b) there must not be any "dark matter."
  $c) the Universe will eventually be entirely cold.
  d) the expansion of the Universe will slow to a halt.

Large Scale Structure and the Cosmic Background Radiation

Lyman-alpha absorbers are clouds of mostly neutral hydrogen believed to be the building blocks of normal galaxies. If anything, we expect their heavy element abundances ("metallicities") should be
  a) larger the farther they are located away from us.
  b) all about the same.
  $c) smaller the farther they are located away from us.
  d) zero, since they have only hydrogen.

Recent observations suggest that galaxies are found in large sheets and filament-like structures, but that there are also enormous holes and voids where no galaxies are found. This shows that the Universe may have
  $a) its largest, most massive structures where filaments join.
  b) formed stars, then galaxies, then clusters of galaxies.
  c) formed clusters of galaxies, then galaxies, then stars.
  d) inflated rapidly, producing sponge-like structure.

The cosmic background radiation provides strong evidence that
  a) star formation has been taking place for billions of years.
  $b) the Universe evolved from a hot, dense state.
  c) colliding galaxies release enormous amounts of synchrotron radiation.
  d) at 3 Kelvin, the early Universe was extremely cool.

The cosmic background radiation is visible in every direction because
  a) we are at the center of the Universe.
  $b) we are looking back to when the Universe was young in every direction.
  c) we are looking back to when the Universe was cool.
  d) it has reflected in every direction over the age of the Universe.

DARK MATTER AND LENSING

Which would be the best evidence for "missing mass" in the Milky Way? The
   a) low density of stars in the spiral arms.
   $b) orbital speeds of gas clouds beyond the orbit of the Sun.
   c) large orbital eccentricities of objects in the disk.
   d) lack of heavy elements in globular cluster stars.

The gravitational effect on light coming from a massive object is that the light will
   a) show no effects from the gravitational field.
   b) travel more slowly through space.
   c) be shifted to higher frequencies.
   $d) have its spectrum shifted to the red.

Which change in appearance of a normal star will occur when a massive dark object (a "MACHO") passes directly between you and the star?
   a) There is no change since dark matter cannot affect light.
   b) The star light disappears momentarily behind the event horizon.
   c) The star dims because the MACHO blocks the light.
   $d) The star brightens temporarily because of gravitational microlensing.

Which of the following measurements can *NOT* be used to measure the amount of dark matter in a cluster of galaxies?
   a) the dispersion in the speeds of galaxies orbiting the cluster center
   $b) the average speed of galaxies orbiting the cluster center
   c) the deflection of light rays from a background object passing by the outskirts of the cluster.
   d) the properties of X-rays emitted by gas that has been heated by falling into the cluster

A solar-mass black hole lies halfway between us and a solar-type star, but not lined up exactly. What do you predict we see when you observe the star?
   a) Bright light from the black hole would outshine the star.
   b) Nothing; the star light would all go down the black hole.
   c) One image of the star, somewhat fainter than the original.
   $d) Two images of the star, one on each side of the black hole.

THE ORIGIN AND EVOLUTION OF LIFE

If a meteor impact killed the dinosaurs, it was most likely due to the resulting
   a) shock wave.
   b) shower of small, high speed fragments.
   c) change in the Earth's orbit, changing the Sun's intensity.
   $d) dust blocking sunlight, killing plants.

You design a weapon that can vaporize any Earth-approaching asteroid. Thank you! How often on average will you have to fire it to protect life on Earth? Approximately every 100
   a) billion years.
 $b) million years.
   c) thousand years.
   d) years.

Viruses have genetic material like plants and animals, but they are not considered to be living because they do not
 $a) metabolize.
   b) breathe.
   c) reproduce.
   d) die.

What important concepts were needed by Darwin to show that evolution could work through natural selection?
   a) catastrophes and floods
   b) cyclic changes
   c) Earth's internal heat
 $d) vast time spans and environmental change

Natural mutations occur in reproduction of life on Earth because
   a) of solar radiation, natural uranium and other carcinogens.
   b) natural amino acids are chemicals that can cause mutations.
 $c) of errors in DNA reproduction in cell nuclei.
   d) sperm and eggs combine genes randomly.

Which effect plays little role in permitting human life to flourish on Earth?
   a) Greenhouse gases lock in warmth.
   b) Jupiter protects Earth from most major comet impacts.
   c) Ozone protects us from ultraviolet radiation.
 $d) The Moon protects Earth from most cosmic rays.

Life as we know it on Earth could only form in a "habitable zone," which is the range of
   a) planet orientations that create seasons.
 $b) distances from a star where most water will be liquid.
   c) latitudes that stay warm during an ice age.
   d) time when there is no longer bombardment by comets and asteroids.

Until the late 1900s, no direct evidence existed for planets around stars other than the Sun. Now that many planets have been detected, we have a better knowledge of what fraction of

a) planets are near to stars.
$b) stars have planets.
c) planets support life.
d) civilizations can detect planets.

You are designing an experiment to detect planets around other stars. Your most effective technique would be
a) direct imaging to see the planet.
b) astrometric detection of the star's orbital motion.
c) detection of radio emissions.
$d) measuring the Doppler shift in the star's spectrum.

You are conducting a search for life outside the Solar System. To maximize your chances of finding planets, you should focus on studying
a) main sequence stars like our Sun.
b) single rather than binary stars.
c) stable rather than pulsating stars.
d) stars at least a few billion years old.
$e) All of these answers are correct.

Classified as a carbonaceous chondrite, the 100 kg Murchison meteorite is suspected to be of cometary origin due to its high water content of 12%. More than 92 different amino acids have been identified in the meteorite, only 19 of which are found on Earth. Which theories about comets might these facts support?
a) They contributed to formation of the oceans and early life on Earth.
b) Cometary material evaporates easily.
c) Comets spend a small fraction of their orbital period near the Sun.
$d) all the above

There are about a trillion stars in the Milky Way Galaxy. Very roughly, how many Earth-like planets are there likely to be in the galaxy?
a) 1
$b) 1 billion
c) 1 million
d) 100 trillion
e) 1000

Living organisms like those on the Earth may exist on planets going around stars other than the Sun because
a) we have detected comets in other stellar systems.
b) life here may have been "planted" by other civilizations.
c) life forms on Earth may have disseminated elsewhere.
$d) the laws of physics and chemistry are universal.
e) some stars emit radio signals.

We might be less likely to find life around an old "Population II" star in the halo of the Milky Way because those stars
    a) are out of the plane of the galaxy.
    b) are too hot and blue.
  $c) may lack heavier elements basic to biological chemistry.
    d) usually pass through the center of the galaxy.

Life is less likely to evolve on planets around massive main sequence stars because massive stars
    a) pull planets inward with their powerful gravity.
    b) engulf planets inside their large radii.
  $c) cease nuclear burning more quickly than required for evolution.
    d) heat planets until they evaporate.

Several physical quantities in the Universe are very finely-tuned to a value that seems to permit our existence. The anthropic principle interprets this, crudely paraphrased, as
    a) "we MUST exist because the Universe has these properties."
  $b) "the Universe must exist because WE do."
    c) "we are alone and unique in the Universe."
    d) "we are at the CENTER of the Universe."

Chapter Six

ASSESSMENT

Classroom assessment of students should prove useful beyond the mechanical production of grades. Good student assessment allows them to adjust their learning strategies, and allows you to react, to adapt, to make midcourse adjustments in course content, pacing, or overall instructional technique. Once the course is over, your teaching can still benefit greatly from student feedback to improve the next class you teach. Finally, having class assessments in hand makes possible meta-assessments, where instructors can benefit from each other's experiences, and the knowledge base expands for educators.

In this chapter, five levels of assessment are briefly discussed.

• *Early diagnostic assessment of the class level.* How to understand students' strengths and weaknesses relative to other classes, so you can adapt lectures, discussions, and assignments accordingly.

• *Interactive assessment.* How ConcepTests and the discussions inherent in Peer Instruction let you gauge class comprehension in real-time.

• *Overall grading.* How within the cooperative learning framework you can assess work by the students — homework, quizzes and exams both inside and outside of class.

• *Student evaluation of your implementation of Peer Instruction for Astronomy.* How well does it work for the students, and how should that be measured?

• *Colleague evaluation of your implementation of Peer Instruction for Astronomy.* What can a colleague see going on from the back of the room that you miss? How can your implementation be made easier or more effective?

• *Cross-institutional evaluation.* How can colleagues, instructors, and educators assess the impact of Peer Instruction on students' appreciation and grasp of astronomy, and guide its use into the future?

Since *Peer Instruction for Astronomy* is meant as an introductory handbook, and an initial database of ConcepTests useful for the classroom, here I merely sketch some of the above issues, and point the way towards further exploration and improvement of the technique in the community of instructors.

EARLY DIAGNOSTIC ASSESSMENT

Peer Instruction is by nature evaluative. The instructor assesses the comprehension level of the class at key points. The students assess their own learning, with the chance for frequent direct comparison with their peers. With consistent use of ConcepTests in the class, assessment becomes a comfortable characteristic of the discourse, rather than a semi-semester peak of punctuated test anxiety and catch-up learning. The feedback provided by Peer Instruction allows the instructor to assess the effectiveness of course content and delivery.

Assessment of the class level may seem intuitive, but there are often unpleasant surprises at exam time. Peer Instruction can help to mitigate those disappointments by keeping the instructor's 'driving force' firmly coupled to the class's "mass." Ideally, ConcepTests should be well-matched to the class level from the outset. "Class level" is a phrase invoking many attributes including preparation and familiarity, conceptual understanding, and problem-solving ability, attention, and enthusiasm. A reliable comparative gauge of the class level is key. How does the class improve over the duration of the course? How does the class level compare to previous classes? To class levels at other institutions? A reliable assessment of class preparation at the start of the course allows you to teach appropriately. Another at the end allows you to evaluate your own methods relative to other semesters, other techniques, and other instructors using the same evaluative tools.

THE ASTRONOMY DIAGNOSTIC TEST

An accurate reading of the class level is important at the outset, so that the course content and pacing can be matched to the class. Teachers normally teach topics that are very familiar to them, and often credit students with more knowledge then they have (McDermott 1991). While acceleration and pacing may vary during the course, you can "lose" students in more ways than one by *starting* at a point that's already beyond them.

The effective use of Peer Instruction should be guided by a firm sense of class level. An instructor new to Peer Instruction can gauge the class level and immediately begin using appropriate-level ConcepTests. The use of a standard gauge has other advantages. If administered both before and after the course, it also provides a quantitative measure of progress achieved during the course. Use of a standard measure is crucial to determine and improve the effectiveness of Peer Instruction relative to other teaching methods. A community-wide database of ConcepTests can contain information about appropriate usage of individual ConcepTests based on class levels ranked with the same instrument.

In physics, the Force Concept Inventory (FCI) has played this role (Hestenes, Wells, and Swackhammer 1992). Pre/post-course test results of thousands of students in 62 introductory mechanics courses have shown that classroom use of interactive engagement methods are on average more than twice as effective as traditional courses in promoting conceptual understanding (Hake 1998). A standard diagnostic test to gauge student understanding is a powerful pedagogical tool, and has been used successfully for college physics courses for the last decade (e.g., Redish and Steinberg 1999). Recently, an Astronomy Diagnostic Test has become available. It is a multiple-choice test, where the distractors are non-trivial responses gleaned from research on common misconceptions.

The Astronomy Diagnostic Test (ADT), is a research-based, multiple-choice assessment of student knowledge and understanding of astronomy at the level of non-majors in an introductory college course. It developed from two prior surveys. The STAR Astronomy Concept Inventory was developed by Philip Sadler (1992) for studying students' ideas about astronomy. The Misconceptions Measure was developed by Michael Zeilik and collaborators (Zeilik 1998). These tests were merged and expanded by instructors in a group called the Collaboration for Astronomy Education Research (CAER). Further refinements removed scientific jargon, limited each question as much as possible to test a single concept, and redesigned many questions so that the correct answer could be determined without reading all of the possible answers. The latest version of the ADT (June 1999) which we present in the **Appendix**, has been administered as a pre-course test to more than 1,500 students in several dozen introductory college astronomy courses. These spanned a range of institutions, including state universities, liberal arts colleges, and community colleges. Many students were also interviewed directly as they answered the ADT questions. Results from these data (Hufnagel et al. 1999) show that the mean pre-course ADT score is similar for all these types of institution, and does not depend on class size. Men showed significantly higher scores than women; 38+/–0.6% vs. 28+/–0.4% for 683 men and 825 women, respectively. Further work incorporating *post-course* results is crucial to understand the significance of these results, and should be used to guide instructors toward more effective teaching methods across the board.

The Astronomy Diagnostic Test is provided in **Appendix II**. At the start of your course, a probe of background knowledge like the ADT is best used for information rather than evaluation. Why intimidate students on the first day with a graded exam? A simple way to avoid that is to ask for students to hand in the results anonymously. Unfortunately, that removes the potential for the diagnostic to be used to group students by background, or to track the progress of individuals over the course duration. You can simply ask for students to hand in the ADT signed, and then tell them that it's not for a grade. In this way, they should have tried their hardest, but those who felt they did poorly will not be unnecessarily discouraged, and students will not be given the impression that the class is points-oriented right from the starting line. Furthermore, after students turn in their individual copies, if you can allocate the class time, have students separate into groups, work together for a few minutes on a group copy of the ADT, and turn that copy in signed for the possibility of bonus group grades. As they hash out their group answers, this should immediately illustrate the utility of Peer Instruction, and engage them early in the process. At the same time, you will have an opportunity to walk around the room and listen in on the discussions for an early assessment of students' understanding, and their ability to function in collaborative learning groups. Before the next class, you can use the individual grades to create heterogeneous longer-term learning groups if you wish, with the option of including other factors like gender, ethnicity, or age.

Another use of an instrument like the ADT is to evaluate how much the students have progressed during the course, so you can offer a modified version just before the final exam. While the level and the concepts should be similar to the original, modifications to the wording or the design of particular realizations of concept testing are encouraged to prevent grades reflecting recall alone. If you are mostly interested in the *gains* that students have made during the course, you might score their results with a gain index (Hake 1998), which effectively removes the variation in absolute levels:

gain index = (%post − %pre) ÷ (100 − %pre)

The gain index can be used to measure an individual students gains for comparison to other students, or you can use it to measure the progress the entire class has made using class means. Use of the standardized ADT will also allow you direct comparison of the class and its progress to similar classes at other institutions, as discussed later. Updates and Spanish language versions are available at

http://solar.physics.montana.edu/aae/adt/
or
http://www.aacc.cc.md.us/scibrhufnagel/

INTERACTIVE CLASS ASSESSMENT

How you grade the ConcepTest solutions used in Peer Instruction affects the degree and style of student participation. Grading students individually may lead to competition that undermines the collaborative nature of the group, unless you take care not to grade on a curve. Assigning all students in a group a single grade based on their group's answer will definitely enliven the group debates. To avoid resentment by the more advanced students, do not make the total grade count for more than a small fraction of the overall class grade.

Some of the better students may resent group grading, and arguably a single group grade can undermine individual accountability, which is crucial to the success of Peer Instruction. One solution is to use both individual and group grades. For example, individual answers to ConcepTests can be used to grade students individually. In a second iteration, students would have the opportunity to counsel their peers toward the answer they think is right. If all students assigned to a given group then record the correct answer, the whole group receives an additional fixed number of bonus points. In this way, both individual accountability and group interdependence are fostered. Even using this method, some students may feel that they are being penalized relative to other groups that have more successful members. A slightly more detailed accounting can alleviate this; *improvement* in a group's overall score also counts as bonus points.

While these methods may sound as if they generate an onerous burden of grading, several techniques can be used to mitigate the extra effort. One option that preserves a balance between group and individual goals is to tally ConcepTest results sometimes for individual scoring and sometimes for a single group score. You might take a quick class tally, then move into a few minutes of Peer Instruction, and have students record their individual answers. Then you can have all the group members sign and turn in a single piece of paper with the group answer on it. Only after you've given them the correct answer do you tell the students whether the ConcepTest is to be scored as individual answer, group answer, or both.

For example, individual ConcepTest responses can be tallied, say using a web interface available to individual students (e.g., on laptops). Students type in their name and password (e.g., a student number), answering A/B/C/D/E to a ConcepTest. The results are tallied, and students break into groups. On the second tally, they again enter their answers, together with the names of other group members. Both individual and bonus points can be awarded according to any desired prescription, and all students are motivated to improve the scores of themselves *and* their peers.

Interactive, real-time web-based grading is discussed at the Galileo web site:

http://galileo.harvard.edu

and at the *Peer Instruction for Astronomy* web site:

http://hea-www.harvard.edu/~pgreen/PIA.html,

and implementation is in progress as of this writing.

READING ASSIGNMENTS
AND READING QUIZZES

Peer Instruction relies heavily on students having read the relevant assigned material before class. Only then will they have the knowledge and insight necessary for fruitful interactions with the instructor (via ConcepTests) and with other students (in their discussion groups). Eric Mazur (1997) presents a short three- or four-question reading quiz at the beginning of each lecture period. These quizzes are counted for bonus points up to a maximum number you deem appropriate. For example, the bonus points can be used to moderate the weight of an exam. You might offer to include bonus points in the lowest exam score the student receives over the duration of the course. Say 8 points have been gained from the reading quizzes, and the student performed worst on the final exam. The final exam is worth 70 points, while the student scored 50. You rescale the final exam score by 62/70 (making 44.3), and add the 8 bonus points back to the resulting exam score (to get 51.3). You can see that the overall effect of bonus points is minor with this method, but students will still be eager to offset their poorest exam score in advance.

Below I include an example of the web form that I use for reading quizzes. The HTML source file for this form is included on the CD-ROM (see Appendix I).

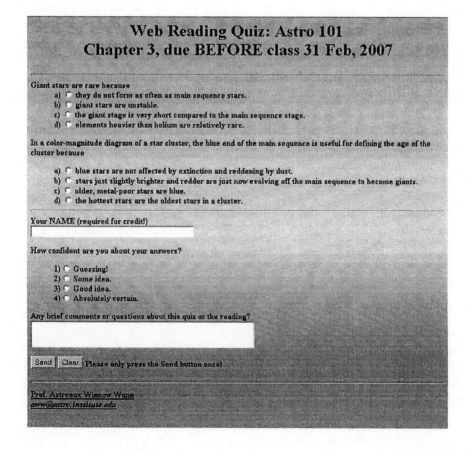

The web form is at a standard web address that I remind students about in class. The quizzes are due before the next class begins, as indicated on the web form. When students fill out the form and submit it, I receive a standard format response that is easy to grade and tally. Below the responses, space is offered for a quick assessment of student confidence, or for comments about the quiz or readings.

As suggested in Chapter Two on Peer Instruction, a useful homework assignment that can play the role of a reading quiz is to have students write their own ConcepTests, based on material from the assigned readings. DaSilva (1995) uses the following method: Students are asked to prepare two high-level conceptual questions, with answers. At the beginning of class, he calls on one of the students to pose the question, and read the answer. Then he spends a couple of minutes discussing the question and its answer with the class. The whole process spans 5–7 minutes. If time permits, another question is solicited from another student. Since each student has written two questions, they cannot avoid the spotlight by claiming that they had the same question. Those not in class, or who have not done the reading, lose the opportunity to earn bonus points. Students are thus motivated to attend prepared. The students' questions provide a daily indicator of their level of commitment and understanding. Judging from their participation in class and their responses on course evaluations, students were generally pleased with the tactic, which gave them more of a feeling of control over their performance than the usual "review your notes, study hard, and you'll get a good grade."

One method allows classroom lessons to be adjusted and organized in response to student submissions "Just-in-Time" for class. Just-in-Time-Teaching (JiTT), described in a book of the same name cited in Chapter Eight on **Readings and Resources**, is a method based on preparatory web assignments that are due a few hours before class. One variation of the JiTT method is well-suited to reading quizzes for Peer Instruction. For details, and numerous examples for Physics, visit the JiTT web site at http://webphysics.iupui.edu/jitt.html. For Peer Instruction, a similar system of web-based reading quizzes has several advantages. A web-based reading quiz motivates students to come to class prepared, and helps them stay organized. It avoids using up class time for reading quizzes, and provides a rapidly tallied measure of student preparedness and understanding to the instructor right before class.

EXAMS

As Eric Mazur (1997) points out in his book for introductory physics classes, Peer Instruction — A User's Manual, "A proper balance between computational and conceptual problems is important." The format and content of examinations should reflect the conceptual underpinnings of your course. Exams often count for a large part of the grade, and so students reasonably tend to gear their study habits toward the exams. If they know early on from the instructor that exams will be at least partly conceptual, similar in philosophy to ConcepTests, then the students will be more attentive and engaged in their efforts to understand the ConcepTests and discussions in class leading up to the exams.

There are several approaches to constructing exam questions. Some are similar to the hints provided for writing ConcepTests. During a lecture, the multiple-choice format of ConcepTests is efficient, facilitating direct comparison between students' responses and a rapid class tally. In the case of exams, you can grade students' answers individually, which leaves room for more open-ended questions that require a variety of skills essential to a successful scientist. Good scientists have to make judgment calls all the time. For instance, when considering a physical phenomenon subject to many influences, we must determine which contributions are important and which can be ignored. In many cases, several assumptions must be made about the physical circumstances. Ballpark estimates of relevant quantities must be made. A model or formula must be developed to represent the problem. Before an answer is derived, the relevant quantities must finally be included appropriately into the model. Below are a couple of simple examples of a back-of-the-envelope conceptual exam questions, with answers provided.

1. The total mass of our Milky Way galaxy is about 10^{12} solar masses. How many solar masses of helium are locked up in stars in our galaxy? Use any assumptions, estimates, or wild guesses you deem necessary, but list them all, and briefly describe your reasoning.

To answer this question completely, students should remember that a large spiral is about 90% dark matter. So only 10^{11} solar masses are in stars. Finally, the helium abundance fraction of stars is about 23% by mass, so about 2.3×10^{10} solar masses of helium is contained in stars in our galaxy. Students should be able to receive *full* credit for answering this question even if they do not recall the exact numbers involved, as long as they correctly include that most of the galaxy is dark (not luminous, stellar) matter, and that stars are mostly hydrogen, not helium.

2. Only about 10% of the total luminosity from a quasar is emitted in the X-ray bandpass, while about 50% is emitted in the ultraviolet. How much longer does a quasar need to be observed in X-rays to get the same number of photons as in the optical?

To answer this question, students must remember that luminosity depends not just on the number of photons, but also on their energy. That is, $L_{opt} = R_{opt} \times E_{opt}$, where R_{opt} is the rate (number per second) of optical photons emitted by the quasar, and E_{opt} is a (representative) energy of each optical photon. Now, since $L_x = R_x \times E_x = (10/50) \times L_{opt}$, we know that $R_x = (1/5) \times R_{opt} \times (E_{opt}/E_x)$. The exposure time required to detect N photons is $T = N/R$. The ratio of X-ray to optical exposure times for the same collecting area is $(T_x/T_{opt}) = (N_x/R_x) \times (R_{opt}/N_{opt})$. Since we want $N_x = N_{opt}$, then $(T_x/T_{opt}) = (R_{opt}/R_x) = 5 \times (E_x/E_{opt})$. Estimate that the energy of an X-ray photon is 1 keV, while the energy of an ultraviolet photon is about 10 eV. So $T_x = 5 \times 100 \times T_{opt}$, or about 500 times longer. This assumes that the collecting area of the two telescopes is the same. Since many students may not state this assumption, it is a good opportunity to point out in class that X-ray telescopes have small collecting areas due to their grazing incidence mirrors, so that exposure times are usually even longer.

OVERALL CLASS GRADING

Assessment of the students during the course is formative (Angelo and Cross 1993), in that it helps measure students' progress towards learning goals. Assessment of students' knowledge and understanding at the end of the unit or course is also valuable as evidence of achievement.

It is up to the instructor to encourage individual responsibility for learning, and to avoid competition. Competition in a cooperative learning environment can discourage students from full participation in sharing their knowledge and teaching their peers. Announcing this in advance to students may bring some surprise, but also relief to them, since the grading will be uncomplicated by the achievements of other students, and since they have more control over their own achievement. The efforts of students to help one another will not harm their own grades, and when they ask for help from others, they are more likely to get it.

If at all possible, do not grade on a curve. Grading on a curve engenders competition, and damages the cooperative learning process. Since students don't know what the level of their peers is until well into the course, it's difficult for them to gauge how hard they should be working.

And if the class is divided into sections, students may wind up being penalized if they sense that they can slack off in a weaker group, while other stronger sections are weighting the mean. An excellent alternative is instead to allow for the accumulation of points from a variety of sources like participation, quizzes, exams, problem sets, or lab work. This makes it easy for students to know exactly where they stand. If they are feeling insecure or just curious, they can tally up their points to date, compare to the maximum, and know immediately where their grade is headed. Simple implementation of a table handed out on the first day makes the system clear to everyone. This is useful to students who can become anxious in a novel learning environment. More detailed tables such as the one below (adopted from Millis and Cottell 1998) can be considered for individual learning units. The table is designed for a hypothetical 5-chapter, 15-lecture course on Extragalactic Astronomy.

COURSE: Extragalactic Astronomy	Possible Points	Actual Earned
Astronomy Diagnostic	0	
Reading Quiz #1	5	
Reading Quiz #2	5	
Reading Quiz #3	5	
Reading Quiz #4	5	
Reading Quiz #5	5	
ConcepTests	15	
Problem Sets		
• Normal Galaxies	10	
• Galaxy Formation & Evolution	10	
• The Big Bang	10	
• Inflationary Cosmology	10	
• Active Galaxies	10	
Midterm Exam	40	
Final Exam	70	
TOTAL POINTS FOR UNIT	200	
FRACTION OF TOTAL		
A>90%; B>80%; C>70%; D>60%; F<60%		

One potential pitfall of cooperative learning is group grades. Students may feel that their grade depends too much on other students, whose diligence and aptitude are beyond their control. Group grades threaten to weaken one of the keystones of cooperative learning, individual accountability. This example outlined in the table is one implementation of grading that involves both individual and group grades, so that both individual achievement and cooperative learning are rewarded and encouraged.

EVALUATING YOUR IMPLEMENTATION

While hundreds of studies effectively document the utility and effectiveness of Peer Instruction and other types of cooperative learning, the content and methods particular to astronomy represent just a tiny fraction of that body of work. More importantly, how Peer Instruction works best for *you*, with your particular course, at your college, has never been addressed before you began to implement it yourself. Your experiences can also be of great benefit to the community of astronomy instructors.

Group Instructional Feedback

A useful classroom assessment of your implementation of Peer Instruction is simply to ask the students for suggestions during the course. This allows you to quickly determine what students find helpful or frustrating, and allow you to adapt your methods if necessary, using student input. Allowing students to suggest what specifically you should start, stop, or continue doing gives them an opportunity to take greater responsibility for the learning experience. Of course you are not obligated to implement their suggestions, but if you are engaged enough to explain the advantages of your chosen methods, then students will likely be more engaged in the group processes, and rank your course higher in their evaluations. Three simple questions to obtain group instructional feedback were suggested by Angelo (1994). Students' written responses should be collected anonymously.

1. "What specific things do I do that really help you learn from group work?
2. What specific things do I do that make it more difficult for you to learn from group work?
3. What are one or two specific, practical changes that we could make to help you learn better in your group?"

These same data can also be collected with a whole-class interview technique. Find a facilitator who has no conflict of interest — perhaps a student from a different class, or an instructor or teaching assistant from a different class or department. You might offer to swap your own time or that of a teaching assistant for the same service. As outlined by Millis and Cottell (1994) in their discussion of Small-Group Instructional Diagnosis, about halfway through the term, the facilitator conducts a 30-minute, in-class interview, either as half of a lecture period, or during lab or discussion sections. Students break up into groups of 6–8, and during about 10 minutes compose a group response that addresses questions similar to those above: (1) List some primary strengths of this course. Give an example or brief explanation of each. (2) List some specific changes that would help you learn more from this course. (3) List other important suggestions for the course, or questions about the course you might have. The facilitator then asks groups to report their results, which are recorded at the front of the class, under heading like "Things to Continue," "Things to Change," and Other Suggestions." Some (indeed, the most important) suggestions may recur, which your facilitator should combine and paraphrase with the group's consensus, and note those most frequently reiterated. When you meet later with the facilitator, you can formulate an action plan to implement those suggestions that are appropriate and reasonable. Identify what elicited each suggestion, and whether it addresses something important to their learning. Is a change feasible and would it be effective? Is the student suggestion the best way to make an effective change? Try to be responsive. Can you effect the change right away? The most challenging part is perhaps the final phase. Since students expectations are raised by the interview, it is important to discuss the results with the students, and explain which suggestions you can address and how.

Course-End Evaluations

Student evaluations provided at the end of the course should help set the agenda for improvements to your next application of Peer Instruction. Students having had the chance to teach each other usually means that their evaluations will be more direct, more constructive, and more relevant. Teachers using Peer Instruction during their lectures receive a wide variety of responses from students, and they are overwhelmingly positive. Some students will continue to feel that the instructor — not the students — should be doing all the teaching. A primary challenge of implementing Peer Instruction in the classroom is to make it both successful at promoting learning, and popular with the students. The key is to provide sufficient direction, and to elicit sufficiently thought-provoking discussions that help students feel that they are truly part of the process. Beyond the standard student evaluation forms that may be provided by your institution, specific suggestions that you solicit at the end of the course will help you to refine your next implementation of Peer Instruction.

Instructor Feedback

Once you have tried Peer Instruction in your lectures, you may also invite colleagues or specialists in faculty development to help assess and improve your implementation of Peer Instruction. The simplest option is to engage a busy colleague whose judgment you respect for an hour or two of their time to observe a lecture and discussions. I recommend seeking student evaluations first, so that you can understand and address any concerns the students may have. Once you have adapted the method as best you can to this particular class, your overall implementation can be eased by the fresh perspective of a colleague who has observed from the back of the classroom. It's a good idea to first outline with them the nature of the course material, what the objectives of the course are, and what to expect from the methodology. You are likely to find that other instructors are pleased by the invitation, and eager to play the constructive critic, especially when it gives them the opportunity to observe and understand a novel lecture mode such as Peer Instruction. Before they come to class, let your colleague know about any specific concerns you have about how to implement Peer Instruction. If you are concerned about how to deal with particular students, or about some aspect of the group dynamics during discussions, let your colleague know so that she can pay particular attention to those issues. Refer her to this book, or other descriptions of collaborative learning techniques from Chapter Eight, **Readings and Resources**. To get the most out of the experience, ask your colleague to provide observational data and positively phrased feedback. Effective feedback will focus on *specific* behaviors that seem to work well or that need modification, rather than on judgments or on abstract theoretical interpretations of the method.

CROSS-INSTITUTIONAL EVALUATION

Education research should guide a natural selection in teaching strategies that engenders real improvement in learning. Ellis and Fouts (1997) and others divide education classroom research into three levels. Level I is theory building; Level II is empirical research, the effect on classes; Level

III is program evaluation, applicable to study of the impact of the strategies on more widespread systems like schools or school districts. We certainly hope that policies influencing these larger systems are based on evidence. While Levels I and II are rather well-developed for Cooperative Learning in general, the technique of Peer Instruction in particular could benefit from more directed feedback, especially as it applies to astronomy. Some meta-analysis of Cooperative Learning was conducted by Qin, Johnson and Johnson (1995). They found that students engaged in Cooperative Learning strategies outperformed their competitive-learning peers in problem-solving areas be they linguistic, non-linguistic, well or poorly defined. Still, Level III research across institutions has barely been touched, since more widespread adoption of the technique is a prerequisite. Your adoption of *Peer Instruction for Astronomy* should mean more fruitful and enjoyable class time for you and your students, but the community of astronomy instructors could also benefit from your experience if you choose to track and record it. Some suggestions for this are supplied here, but will be expanded as more instructors participate. Keep track of the web site for further progress in evaluation, and simple opportunities to contribute.

Chapter Seven

EPILOG

Try Peer Instruction in your college Introductory Astronomy class. It's not hard to implement, and yields rapid rewards for both you and your students. While Peer Instruction is scalable to your level of interest and commitment, you will benefit by putting much more than a toe in the water. Experience shows that a full implementation, meaning two or three ConcepTests and discussions per lecture hour, accomplishes a lot. *Peer Instruction for Astronomy,* thoughtfully administered, will almost surely

1. raise class attendance and lower course attrition.
2. boost and hold the interest of your students.
3. heighten your awareness of students' comprehension.
4. highlight common misconceptions to be addressed directly in lecture.
5. increase student understanding of key physical concepts.
6. improve student retention.
7. develop students' ability to communicate scientific ideas.
8. enhance students' collaborative skills.
9. raise student satisfaction with your course and appreciation of your teaching.

Now, why should you buy all that? Although there is a huge body of research documenting the effectiveness of cooperative learning techniques like Peer Instruction, the effectiveness of the specific techniques discussed here, and their particular application to astronomy have only begun to be studied. Are all the above points true? In what circumstances? How can you trouble-shoot your implementation of *Peer Instruction for Astronomy*? It is crucial that the experience of teachers like yourself be shared in the community of astronomy instructors and educators. *Peer Instruction for Astronomy* should be researched and documented in detail. Don't just go it alone. Check the following web sites, readings, and references in Chapter Eight so that we can all learn and benefit from each other. Share the wealth!

Chapter Eight

READINGS AND RESOURCES

Hundreds of worthwhile articles and texts discuss Cooperative Learning and (more recently) Peer Instruction in higher education. Some are theoretical, but many case studies are available in a variety of disciplines, as are studies based on larger samples. Millis and Cottell (1998) is especially helpful for a general description and overview of Cooperative Learning, and they lay out many possible methods of application. Since more are being published every day, the readings provided below represent a good start rather than an exhaustive list. Explore for yourself!

READINGS

Bligh, Donald A. 2000. *What's the Use of Lectures?* San Francisco: Jossey-Bass Publishers.

Christian, W., and Belloni, M. 2001. *Physlets: Teaching Physics with Interactive Curricular Material.* Upper Saddle River, NJ: Prentice Hall.

Davis, James R. 1993. *Better Teaching, More Learning: Strategies for Success in Postsecondary Settings*. American Council on Education, Oryx Press; Phoenix, AZ.

Dunkin, Michael J., and Barnes, Jennifer. 1986. "Research on Teaching in Higher Education" and Walberg, Herbert J. "Synthesis of Research on Teaching," in *Handbook of Research on Teaching* (3rd edition), edited by Merlin C. Wittock. New York: Macmillan.

Ellis, Arthur K., and Fouts, Jeffrey T. 1997. *Research on Educational Innovations* (2nd edition). Larchmont, NY: Eye on Education.

Gokhale, A. A. 1995. "Collaborative Learning Enhances Critical Thinking," *Journal of Technology Education*, 7, 1.

Hake, R. 1998. "Interactive Engagement Versus Traditional Methods: a Six-Thousand Student Survey of Mechanics Test Data for Introductory Physics Courses," *Am. J. Phys.* 66, 64.

Johnson, D. W., Johnson, R. T., and Smith, K. A. 1991. "Cooperative Learning: Increasing College Faculty Instructional Productivity." *ASHE-ERIC Higher Education Report* No. 4. Washington, D.C.: The George Washington University, School of Education and Human Development.

_____. 1996. "Academic Controversy: Enriching College Instruction through Intellectual Conflict." *ASHE-ERIC Higher Education Report Volume* 25 No. 3. Washington, D.C.: The George Washington University, School of Education and Human Development.

_____. 1991. *Active Learning: Cooperation in the College Classroom.* Edina, MN: Interaction Book Co.

Johnson, R. T., and Johnson, D. W. 1986. "Action research: Cooperative learning in the science classroom." *Science and Children*, 24, 31–32.

Mazur, E. 1991. *Peer Instruction: A User's Manual.* Upper Saddle River, NJ: Prentice Hall.

McDermott, L. C. 1991. "Millikan Lecture: What We Teach and What Is Learned — Closing the Gap," *American Journal of Physics*, 59, 301–315.

McDermott, L. ,and Shaffer, P. 2002. *Tutorials in Introductory Physics.* Upper Saddle River, NJ: Prentice Hall.

Millis, Barbara J., and Cottell, Philip G., Jr. 1998. *Cooperative Learning for Higher Education Faculty.* American Council on Education, Oryx Press; Phoenix, AZ.

Novak, G., Patterson, E., Gavrin, A., and Christian, W. 1999. *Just-in-Time Teaching: Blending Active Learning with Web Technology,* Upper Saddle River, NJ: Prentice Hall.

O'Kuma, T., Maloney, D., and Hieggelke, C. 2000. *Ranking Task Exercises in Physics.* Upper Saddle River, NJ: Prentice Hall.

Rau, W., and Heyl, B. S. 1990. "Humanizing the College Classroom: Collaborative Learning and Social Organization Among Students." *Teaching Sociology*, 18, 141–155.

Slavin, R. E. 1989. "Research on Cooperative Learning: An International Perspective." *Scandinavian Journal of Educational Research*, 33(4), 231.

Tobias, S. 1992. *Revitalizing Undergraduate Science: Why Some Things Work and Most Don't*, Tucson, AZ: Research Corporation.

Totten, S., Sills, T., Digby, A., and Russ, P. 1991. *Cooperative Learning: A Guide to Research.* New York: Garland.

Sadler, P. M. 1987. "Misconceptions in Astronomy." In *2nd International Seminar on Misconception and Educational Strategies in Science and Mathematics in Ithaca, NY*, ed. Joseph D. Novak. Ithaca, NY: Cornell University Press.

Springer, L., Stanne, M. E., and Donovan, S. S. 1999. "Effects of Small-Group Learning on Undergraduates in Science, Mathematics, Engineering, and Technology: A Meta-Analysis." *National Institute for Science Education, Research Monograph* No. 11; University of Wisconsin-Madison. Published in Review of Educational Research, Volume 69(1), 21.

Penner, J. G. 1984. *Why Many College Teachers Cannot Lecture.* Springfield, IL: Charles C. Thomas.

Zeilik, M. et al. 1997. "Conceptual Astronomy: A novel model for teaching postsecondary science courses," *American Journal of Physics, 65,* 987.

WEB RESOURCES

The sheer volume of information, and the many sites resounding with false promise mean the web can be overwhelming, so a quick list of some relevant sites, each with a brief description, can ease your way. With the incredible growth and change on web sites, any such list must be both incomplete and quickly obsolete. While I highly recommend the sites below, I cannot endorse any particular products or methods you see there. My emphasis is on free sites and free software. I hope that the list saves you some time. For web novices (if any such folk are left), a quick jumpstart to finding such sites on your own would be to access your favorite web search tool like Google, and use keywords and phrases such as cooperative or collaborative learning, and Peer Instruction.

http://aer.noao.edu
The **Astronomy Education Review**, hosted by the National Optical Astronomy Observatories, has articles on a broad range astronomy and space science education topics, ideas for innovative teaching and outreach, funding opportunities, meeting announcements, etc.

http://www.astrosociety.org/education/resources/resources.html
The **Astronomical Society of the Pacific**'s Education Resource page, authored in large part by Andy Fraknoi, lists and links astronomy education projects and resources in the United States, and provides publications, some free.

http://solar.physics.montana.edu/aae/adt/   or
http://www.aacc.cc.md.us/scibrhufnagel/
The **Astronomy Diagnostic Test** is provided in Appendix II of this book, but updates, statistics of usage, and a Spanish language version are provided at these web sites.

http://www.le.ac.uk/cc/ltg/castle/
The **CASTLE** toolkit has been developed so that tutors and course managers can create online interactive multiple-choice questions (MCQs) quickly and easily without any prior knowledge of HTML, cgi, or similar scripting languages. The resulting tests can be saved as html documents.

http://www.cat.ilstu.edu/teaching_tips/collab.shtml
The **Center for the Advancement of Teaching** at Illinois State University provides a list of links related to collaborative/cooperative learning.

http://www.stedwards.edu/cte/resources.htm
The **Center for Teaching Excellence** at St. Edwards University has compiled a list of links and references for information on active and cooperative learning, group study, assessing learning outcomes, and more.

http://wings.buffalo.edu/vpaa/ctlr/
The **Center for Teaching and Learning Resources** at SUNY Buffalo highlights teaching strategies and faculty development activities.

http://shiraz.as.arizona.edu/
The **Conceptual Astronomy and Physics Education Research** (CAPER) team's mission is to develop and disseminate effective instructional

interventions and authentic assessment strategies based on research in student understanding, with a focus on Collaborative Learning strategies. The team conducts research and public outreach activities in the areas of physics, astronomy, and earth/space science.

http://www.clcrc.com
The **Cooperative Learning Center** at the University of Minnesota helps foster effective student interactions in a wide range of cooperative learning scenarios with these links and information from well-known experts Roger and David Johnson.

http://www.le.ac.uk/cc/ltg/castle/resources/mcqman/mcqappb.html
Designing and Managing Multiple Choice Questions, including dos and don'ts of writing them, written by John Carneson, Georges Delpierre, and Ken Masters at the University of Capetown, South Africa.

http://www.iasce.net
The **International Association for the Study of Cooperative Education** is a non-profit group for educators who researcher and practice Cooperative Learning. The IASCE supports the development and dissemination of CL-related research.

http://www.jigsaw.org
Jigsaw Classroom is a cooperative learning technique that emphasizes individual responsibility, and optimizes cooperation of mixed groups for medium- to longer-term projects like compiling reports.

http://webphysics.iupui.edu/jitt/jitt.html
Just-in-Time Teaching (JiTT) is a teaching and learning strategy comprised of two elements: classroom activities that promote active learning and World Wide Web resources that are used to enhance the classroom component. Students respond electronically to carefully constructed web-based assignments that are due a few hours before class, and the instructor reads the student submissions "just-in-time" to adjust the lesson content and activities to suit the students' needs. The web site briefly describes the method, provides links and references the Prentice-Hall book *Just-in-Time Teaching* (see **Readings**).

http://www.tlzinc.com
The **Learning Zone** highlights books by Michael Zeilik that provide classroom-tested astronomy activities focussing on cooperative learning.

http://www.geocities.com/~maacie
The **Mid-Atlantic Association for Cooperation in Education** offers Cooperative Learning tips and ideas from the archives of the MAACIE Newsletter, and links to other sites on Cooperative Learning in education and information about other Cooperative Learning resources.

http://www.wcer.wisc.edu/nise/cl1
The **National Institute for Science Education, College Level One** site provides several major resources on (1) Collaborative Learning (2) Field-Tested Learning Assessment Guide (FLAG), and the Student Assessment of their Learning Gains (SALG), and (3) Learning through Technology.

http://www.ntlf.com/html/lib/bib/bib.htm
The **National Teaching and Learning Forum** online provides a bibliography of selected resource materials, and links, with some full-text versions of published books available online.

http://hea-www.harvard.edu/~pgreen/PIA.html
At this *Peer Instruction for Astronomy* web site, the ConcepTests Library from Chapter Five remains accessible on the web where Astronomy instructors can both access and contribute. Feedback from instructors on the content and scoring of individual ConcepTests will be used to continually adapt and refine the Library in the future, making it a dynamic, accessible tool suitable for direct use in the classroom, but also as a potential database for research on and assessment of the technique of Peer Instruction and its results.

http://mazur-www.harvard.edu/education/pi.html
Peer Instruction for Physics by Eric Mazur's group at Harvard explores collaborative learning in large lectures. The related site from the Galileo project http://galileo.harvard.edu/home.html includes a growing, searchable database of ConcepTests for physics and other sciences. Mazur's method is described in *Peer Instruction: A User's Manual* (see **Readings**), where physics ConcepTests are provided. The book also includes two nationally recognized tools for evaluation, the Force Concept Inventory and the Mechanics Baseline Test, usable as pre-tests and post-tests to evaluate both teaching effectiveness and student learning. Reading quizzes, conceptual exam questions, and ConcepTests intended for a one-year introductory college physics course are included.

http://quiz.4teachers.org
Quizstar allows you to create a custom quiz on their web site that students can take online. The quizzes are stored by a name you choose at the Quizstar site, and can be activated only for chosen dates and times.

http://ase.tufts.edu/cte/pages/resource.htm
The **Tufts Center for Academic Excellence**, like many such sites at colleges around the country, provides Resources on teaching and Learning.

http://www.uoregon.edu/~tep/library/articles/
The **Teaching Effectiveness Program** at the University of Oregon provides articles and references relevant to collaborative/cooperative learning.

http://www.uga.berkeley.edu/sled/bgd/collaborative.html
From **Tools for Teaching** by Barbara Gross Davis; Jossey-Bass Publishers: San Francisco 1993. This page describes design and implementation of group study in the classroom.

http://webct2_2.prenhall.com/public/walker_phy1vol1/
This WebCT course has been designed to be an online companion to the textbook **Physics**, Volume I, by James S. Walker, offering a diverse array of web tools designed to help instructors teach more effectively and students to learn and understand the material in order to apply it in a broad context.

http://www.webct.com
WebCt is a commercial software company that offers course management that must be purchased and the user licensed. While many of the tools are for building web-based course content, several provide for interactive, real-time classroom assessment, self-tests and surveys appropriate to Peer Instruction.

http://www.extropia.com/scripts/multiple_choice.html
WebExam is free downloadable software that allows you to create your own multiple-choice exams on the Web. If you create an answer key database, it will grade the answers submitted by a user and gather all completed exams in a database

http://webphysics.davidson.edu/

The **WebPhysics** project was started by Wolfgang Christian and Gregor Novak during the spring of 1995 to facilitate Teaching and Learning Physics with World Wide Web Technology. Web Physics is an outlet for small volume HTML-based curricular material produced by physics instructors around the country.

http://fpg.uwaterloo.ca/WEBTEST/

WEBTEST can be used to create an interactive testing, tutorial or survey environment. The software is free, is written in Perl and intended for UNIX.

http://chemeng.mcmaster.ca/pbl/pbl.html

This site works on skill building for "**Problem-Based Learning**" that emphasizes group-based problem solving. *PBL* is any learning environment in which the problem drives the learning. That is, before students learn some knowledge, they are given a problem. The problem is posed so that students discover that they need to learn something new before they can solve it.

REFERENCES

AIP Education and Employment Statistics Division. 1995. "Skills Used Frequently by Physics Bachelors in Selected Employment Sectors."

Angelo, T. A., and Cross, K. P. 1993. *Classroom Assessment Techniques: A Handbook for College Teachers* (2nd edition). San Francisco: Jossey-Bass.

Angelo, T. A. 1994, Spring. "Using Assessment to Improve Cooperative Learning." *Cooperative Learning and College Teaching,* 4(3), 5.

Atwood, R. K., and Atwood, V. A. 1996. "Preservice Elementary Teachers' Conceptions of the Causes of Seasons." *Journal of Research in Science Teaching*, 33(5), 553.

Beckman, M. 1990. "Collaborative Learning: Preparation for the Workplace and Democracy." *College Teaching*, 38(4), 128.

Bonwell, C. C. and Sutherland, T. E. 1996. "The Active Learning Continuum: Choosing Activities to Engage Students in the Classroom." In T. E. Sutherland, and C. C. Bonwell (eds.), *Using Active Learning in College Classes: A Range of Options for Faculty* (p. 3*)*. Cited in *New Directions for Teaching and Learning,* n° 67. San Francisco: Jossey-Bass.

Burns, M. 1993. "Teaching What to Do in Arithmetic vs. Teaching What To Do and Why." *Educational Leadership*, 43, 34. Cited in James R. Davis, *Better Teaching, More Learning: Strategies for Success in Postsecondary Settings.* Phoenix, AZ: Oryx Press.

Carnevale, A. P., and Fry, R. 2000. *Crossing the Great Divide: Can We Achieve Equity When Generation Y Goes to College?* Princeton, NJ: Educational Testing Service (ETS) Reports.

Carpenter, T. P., Lindquist, M. M., Matthews, W., and Silver, E. A. 1983, "Results of the Third NAEP Mathematics Assessment: Secondary School." *Mathematics Teacher*, 76, 652.

Christian, W., and Titus, A. 1998. "Developing Web-Based Curricula Using Java Applets." *Computers in Physics,* 12, 227.

Clement, J., Brown, D., and Zeitsman, A. 1989. "Not All Preconceptions Are Misconceptions: Finding 'Anchoring Conceptions' for Grounding Instruction on Students' Intuitions." *International Journal of Science Education*, 11, 554.

Comins, N. 2001. *Heavenly Errors: Misconceptions About the Real Nature of the Universe.* New York: Columbia University Press.

Comins, N. 2000. "Astronomy Education Research: Elements of Student Learning (with Important Contributions by a Astrononomy 101 Panel). *AAS Education Report*, available at http://www.aas.org/education/handouts/1_2000/learnissues.pdf.

Cooper, J., and Associates. 1990. *Cooperative Learning and College Instruction.* Long Beach: Institute for Teaching and Learning, California State University.

Costin, F. 1972. "Lecturing Versus Other Methods of Teaching: A Review of Research." *British Journal of Education Technology*, 3(1), 4.

Davidson, N. 1990. "The Small-Group Discovery Method in Secondary- and College-level Mathematics." In N. Davidson (ed.), *Cooperative Learning in Mathematics: A Handbook for Teachers,* p 335. Menlo Park, CA: Addison-Wesley.

Fraknoi, A. 1998. "Astronomy Education in the United States". Invited talk given at the 189th Meeting of the American Astronomical Society, Toronto, Canada, cited from http://www.astrosociety.org/education/resources/useduc.html

Gardner, A. L., Mason, C. L., and Matyas, M. L. 1989. "Equity, Excellence, and 'Just Plain Good Teaching.' " *The American Biology Teacher*, 51(2), 72.

Goldsmid, C. A., and Wilson, E. K. 1980. *Passing On Sociology.* Washington, D.C.: American Sociological Association Teaching Resources Center.

Goodsell, A., Maher, M., Tinto, V, and Associates (eds.). 1992. *Collaborative Learning: A Sourcebook or Higher Education.* University

Park, PA: National Center on Postsecondary Teaching, Learning, and Assessment, Pennsylvania State University.

Hammer, D. 2000. "Student Resources for Learning Introductory Physics." *American Journal of Physics, Physics Education Research Supplement*, 68 (S1), S52–S59.

Hestenes, D., Wells, M., and Swackhammer, G. 1992. "Force Concept Inventory." *Phys. Teach.*, 30, 14.

Hufnagel, B., Slater, T., Deming, G., Adams, J., Adrian, R. L., Brick, C., and Zeilik, M. 1999. "Pre-Course Results from the Astronomy Diagnostic Test." *PASA*, 17, 2.

Johnson, D. W., Johnson, R. T, and Holubec, E. 1990. *Circles of Learning: Cooperation in the Classroom*. Edina, MN: Interaction Book Co.

Johnson, D.W., and Johnson, R.T. 1991. *Learning Together and Alone: Cooperative, Competitive, and Individualistic* (3d edition). Upper Saddle River, NJ: Prentice Hall.

Johnson, D. W., Johnson, R. T., and Smith, K. A. 1991a. *Cooperative Learning: Increasing College Faculty Instructional Productivity*. ASHE-ERIC Higher Education Report No. 4. Washington, D.C.: School of Education and Human Development, George Washington University.

_____. 1991b, In *Active Learning: Cooperation in the College Classroom*. Edina, MN: Interaction Book Company.

Lightman, A. P., Miller, J. D., and Leadbeater, B. J. 1987. "Contemporary Cosmological Beliefs." In J. D. Novak (ed.), *Proceedings of the second international seminar on misconceptions and educational strategies in science and mathematics, Vol. III* (pp. 309–321). Ithaca, NY: Department of Education, Cornell University.

Mathews, R. S., Cooper, J. L., Davidson, N., and Hawkes, P. 1995. "Building Bridges Between Cooperative and Collaborative Learning." *Change: The Magazine of Higher Learning*, Jul/Aug, 35.

Meyers, C., and Jones, T. B. 1993. *Promoting Active Learning. Strategies for the College Classroom*. San Francisco: Jossey-Bass.

Millis, B. J., and Cottell, P. G. 1998. *Cooperative Learning for Higher Education Faculty,* Phoenix, AZ: Oryx Press.

Mulvey, P. J., and Nicholson, S. 2001. "Enrollments and Degrees Report," AIP Education and Employment Statistics Division, AIP Pub. #R-151.37, available at http://www.aip.org/statistics/trends/reports/ed.pdf

Qin, Z., Johnson, D. W., and Johnson, R. T. 1995. "Cooperative Versus Competitive Efforts and Problem Solving." *Review of Educational Research,* 65(2), 129.

Piaget, J. 1950. *The Psychology of Intelligence.* New York: Harcourt.

Redish, E. G., and Steinberg, R. N. 1999. *Physics Today,* 52, 1.

Sadler, P. 1992. *The Initial Knowledge State of High School Astronomy Students* (Ed.D. Dissertation, Harvard School of Education).

_____. 1998. "Psychometric Models of Student Conceptions in Science" *Journal of Research in Science Teaching,* 35(3), 265.

Schlatter, M. 2001. "Writing ConcepTests for a Multivariable Calculus Course." *PRIMUS,* submitted.

Schneps, M. H. 1987. *A Private Universe.* Available from Pyramid Films and Video, 2801 Colorado Avenue, Santa Monica, CA 90404.

Seymour, E. 1993. "Why Are the Women Leaving?" Lecture, NECUSE Conference. Brown University, April 10, 1993, as cited in "A Guide for Faculty," The New England Consortium for Undergraduate Science Education 1996, at http://www.brown.edu/Administration/Dean_of_the_College/homepginfo/equity/Equity_handbook.html

Silva, F. 1995. "Student-Generated Test Questions." *Teaching Resources Center,* Indiana University, 7(2).

Subrahmanyam, K., Kraut, R. E., Greenfield, P. M., and Gross, E. F. "The Impact of Home Computer Use on Children's Activities and Development." *Children and Computer Technology,* 10(2).

Sutherland, T. E., and Bonwell, C. C., Eds. 1996. "Using Active Learning in College Classes: A Range of Options for Faculty" in *New Directions for Teaching and Learning*, n° 67. San Francisco: Jossey-Bass.

Wilkerson, L. 1994. "Identification of Skills for the Problem-Based Tutor: Student and Faculty Perspectives" (seminar). McMaster University, Hamilton, ON.

Zeilik, M., et al. 1997. "Conceptual Astronomy: A Novel Model for Teaching Postsecondary Science Courses," *American Journal of Physics*, 65, 987.

Zeilik, M., Schau, C., and Mattern, N. 1998. "Misconceptions and Their Change in University-Level Astronomy Courses." *The Physics Teacher*, 36 (2), 104.

Zeilik, M., Shau, C., and Mattern, N. 1999. "Conceptual Astronomy. II. Replicating Conceptual Gains, Probing Attitude Changes Across Three Semesters." *American Journal of Physics*, 67(10), 923.

APPENDIX I

CD-ROM INSTRUCTIONS

The CD-ROM at the back of *Peer Instruction for Astronomy* contains all of the ConcepTests from Chapter Five in two different formats. A single Microsoft Word document allows searching and editing for people who prefer that format. A single Adobe PDF (Portable Document Format) file is also included, with bookmarks for each of the 25 named ConcepTest Library chapter sections, as follows:

THE NIGHT SKY
MEASURES AND METHODS
TELESCOPES
HISTORY
GENERAL MOTION/FORCES
SCALES OF SIZE, DISTANCE, MASS, AND POWER
THE ELEMENTS
RADIATION AND THE ELECTROMAGNETIC
  SPECTRUM
THE EARTH
EARTH AND MOON
THE MOON
THE PLANETS
ASTEROID, METEORS, AND COMETS
THE SUN
BASIC STELLAR PROPERTIES
STAR FORMATION
ENERGY GENERATION IN STARS AND STELLAR
  EVOLUTION
BINARY STAR SYSTEMS
STELLAR POPULATIONS
OUR GALAXY
NORMAL GALAXIES
ACTIVE GALAXIES AND QUASARS
COSMOLOGY
DARK MATTER AND LENSING
THE ORIGIN AND EVOLUTION OF LIFE

Within the PDF document, each ConcepTest is on a single page for easy printing and classroom display, for instance as transparencies. PDF files can be read with Adobe Acrobat Reader. Your computer may have Acrobat pre-installed, but it is free software easily downloaded from Adobe's site http://www.adobe.com which provides simple instructions for download and use.

If you are unable to access the material on the disks, please contact me or Prentice Hall by writing to Physics Editor, Higher Education Division, Prentice Hall, 1 Lake Street, Upper Saddle River, NJ 07458.

Note that virtually all of the Conceptests from Chapter Five are reproduced in a growing and evolving ConcepTest Library at

http://hea-www.harvard.edu/~pgreen/PIA.html

APPENDIX II:

The Astronomy Diagnostic Test

The ADT was designed for undergraduate, non-science majors taking their first astronomy course, and was developed by the Collaboration for Astronomy Education Research (CAER) including Jeff Adams, Rebecca Lindell Adrian, Christine Brick, Gina Brissenden, Grace Deming, Beth Hufnagel, Tim Slater, and Michael Zeilik. The first 21 questions are the content portion of the test, and the final 12 questions collect demographic information.

Introductory Astronomy Survey

1. As seen from your current location, when will an upright flagpole cast no shadow because the Sun is directly above the flagpole?

   A. Every day at noon.
   B. Only on the first day of summer.
   C. Only on the first day of winter.
   D. On both the first days of spring and fall.
   E. Never from your current location.

2. When the Moon appears to completely cover the Sun (an eclipse), the Moon must be at which phase?

   A. Full
   B. New
   C. First quarter
   D. Last quarter
   E. At no particular phase

3. Imagine that you are building a scale model of the Earth and the Moon. You are going to use a 12-inch basketball to represent the Earth and a 3-inch tennis ball to represent the Moon. To maintain the proper distance scale, about how far from the surface of the basketball should the tennis ball be placed?

   A. 4 inches (1/3 foot)
   B. 6 inches (1/2 foot)

C. 36 inches (3 feet)
D. 30 feet
E. 300 feet

4. You have two balls of equal size and smoothness, and you can ignore air resistance. One is heavy, the other much lighter. You hold one in each hand at the same height above the ground. You release them at the same time. What will happen?

A. The heavier one will hit the ground first.
B. They will hit the ground at the same time.
C. The lighter one will hit the ground first.

5. How does the speed of radio waves compare to the speed of visible light?

A. Radio waves are much slower.
B. They both travel at the same speed.
C. Radio waves are much faster.

6. Astronauts inside the Space Shuttle float around as it orbits the Earth because

A. there is no gravity in space.
B. they are falling in the same way as the Space Shuttle.
C. they are above the Earth's atmosphere.
D. there is less gravity inside the Space Shuttle.
E. more than one of the above.

7. Imagine that the Earth's orbit were changed to be a perfect circle about the Sun so that the distance to the Sun never changed. How would this affect the seasons?

A. We would no longer experience a difference between the seasons.
B. We would still experience seasons, but the difference would be much LESS noticeable.
C. We would still experience seasons, but the difference would be much MORE noticeable.
D. We would continue to experience seasons in the same way we do now.

8. Where does the Sun's energy come from?

   A. The combining of light elements into heavier elements
   B. The breaking apart of heavy elements into lighter ones
   C. The glow from molten rocks
   D. Heat left over from the Big Bang

9. On about September 22, the Sun sets directly to the west as shown on the diagram below. Where would the Sun appear to set two weeks later?

   A. Farther south
   B. In the same place
   C. Farther north

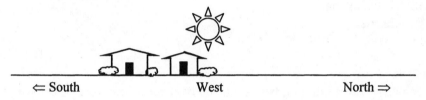

   ⇐ South   West   North ⇒

10. If you could see stars during the day, this is what the sky would look like at noon on a given day. The Sun is near the stars of the constellation Gemini. Near which constellation would you expect the Sun to be located at sunset?

   A. Leo
   B. Cancer
   C. Gemini
   D. Taurus
   E. Pisces

Sun

Gemini

Cancer

Taurus

Leo

Pisces

   ⇐ East   South   West

11. Compared to the distance to the Moon, how far away is the Space Shuttle (when in space) from the Earth?

    A. Very close to the Earth
    B. About half way to the Moon
    C. Very close to the Moon
    D. About twice as far as the Moon

12. As viewed from our location, the stars of the Big Dipper can be connected with imaginary lines to form the shape of a pot with a curved handle. To where would you have to travel to first observe a considerable change in the shape formed by these stars?

    A. Across the country
    B. A distant star
    C. Europe
    D. Moon
    E. Pluto

13. Which of the following lists is correctly arranged in order of closest-to-most-distant from the Earth?

    A. Stars, Moon, Sun, Pluto
    B. Sun, Moon, Pluto, stars
    C. Moon, Sun, Pluto, stars
    D. Moon, Sun, stars, Pluto
    E. Moon, Pluto, Sun, stars

14. Which of the following would make you weigh half as much as you do right now?

    A. Take away half of the Earth's atmosphere.
    B. Double the distance between the Sun and the Earth.
    C. Make the Earth spin half as fast.
    D. Take away half of the Earth's mass.
    E. More than one of the above

15. A person is reading a newspaper while standing 5 feet away from a table that has on it an unshaded 100-watt light bulb. Imagine that the table were moved to a distance of 10 feet. How many light bulbs in

total would have to be placed on the table to light up the newspaper to the same amount of brightness as before?

A. One bulb.
B. Two bulbs.
C. Three bulbs.
D. Four bulbs.
E. More than four bulbs.

16. According to modern ideas and observations, what can be said about the location of the center of the Universe?

A. The Earth is at the center.
B. The Sun is at the center.
C. The Milky Way Galaxy is at the center.
D. An unknown, distant galaxy is at the center.
E. The Universe does not have a center.

17. The hottest stars are what color?

A. Blue
B. Orange
C. Red
D. White
E. Yellow

18. The diagram below shows the Earth and Sun as well as five different possible positions for the Moon. Which position of the Moon would cause it to appear like the picture at right when viewed from Earth?

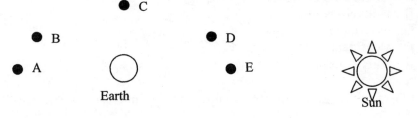

19. You observe a full Moon rising in the east. How will it appear in six hours?

A.　　　　　　B.　　　　　　C.　　　　　　D.

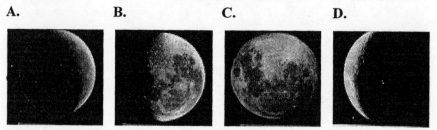

Images © UC Regents/Lick Observatory. Unauthorized use prohibited.

20. With your arm held straight, your thumb is just wide enough to cover up the Sun. If you were on Saturn, which is 10 times farther from the Sun than the Earth is, what object could you use to just cover up the Sun?

    A. Your wrist
    B. Your thumb
    C. A pencil
    D. A strand of spaghetti
    E. A hair

21. Global warming is thought to be caused by the

    A. destruction of the ozone layer.
    B. trapping of heat by nitrogen.
    C. addition of carbon dioxide.

22. In general, how confident are you that your answers to this survey are correct?

    A. Not at all confident (just guessing)
    B. Not very confident
    C. Not sure
    D. Confident
    E. Very confident

23. What is your college major (or current area of interest if undecided)?

    A. Business

B. Education
C. Humanities, Social Sciences, or the Arts
D. Science, Engineering, or Architecture
E. Other

24. What was the last math class you completed prior to taking this course?

A. Algebra
B. Trigonometry
C. Geometry
D. Pre-Calculus
E. Calculus

25. What is your age?

A. 0-20 years old
B. 21-23 years old
C. 24-30 years old
D. 31 or older
E. Decline to answer

26. Which best describes your home community (where you attended high school)?

A. Rural
B. Small town
C. Suburban
D. Urban
E. Not in the USA

27. What is your gender?

A. Female
B. Male
C. Decline to answer

28. Which best describes your ethnic background?

A. African-American
B. Asian-American

    C. Native-American
    D. Hispanic-American
    E. None of the above (see question 29 below)

29. Which best describes your ethnic background?

    A. African (not American)
    B. Asian (not American)
    C. White, non-Hispanic
    D. Multicultural
    E. None of the above (see question 28 above)

30. How good at math are you?

    A. Very poor
    B. Poor
    C. Average
    D. Good
    E. Very good

31. How good at science are you?

    A. Very poor
    B. Poor
    C. Average
    D. Good
    E. Very good

32. Which best describes the level of difficulty you expect/experienced from this course?

    A. Extremely difficult for me
    B. Difficult for me
    C. Unsure
    D. Easy for me
    E. Very easy for me

33. How many astronomy courses at the college level have you taken?

    A. I'm re-taking this course.
    B. This is my first college-level astronomy course.

C. This is my second college-level astronomy course.
D. I've completed more than two other college-level astronomy courses.

END OF SURVEY

Index

Minimum System Requirements

Windows/PC
Pentium processor
Windows 98/ME/NT 4.x/2000/XP
32 MB hard disk space
64 MB of RAM
4X CD-ROM (12X or higher recommended)
Microsoft Word, version Office 97 or greater (or compatible software) to view and edit the ConcepTest.doc
Adobe Acrobat Reader 5.0 © 2002 or above is required to view and print the pdfs
Netscape Navigator 4.08 © 1995-2002 or above to view the Flashcard JPegs
(Installers for Acrobat Reader 5.0 and Netscape 4.08 included on the CD ROM)

Macintosh
PowerPC 300 MHz processor or higher
OS 8/9/X
32 MB hard disk space
64 MB of RAM
4X CD-ROM (12X or higher recommended)
Microsoft Word, version Office 97 or greater (or compatible software) to view and edit the ConcepTest.doc
Adobe Acrobat Reader 5.0 © 2002 or above is required to view and print the pdfs
Netscape Navigator 4.08 © 1995-2002 or above to view the Flashcard JPegs
(Installers for Acrobat Reader 5.0 and Netscape 4.08 included on the CD ROM)